NUREG–1552, Supp. 1

I0488452

Fire Barrier Penetration Seals in Nuclear Power Plants

Manuscript Completed: December 1998
Date Published: January 1999

C. S. Bajwa, K. S. West

Division of Systems Safety Analysis
Office of Nuclear Reactor Regulation
U.S. Nuclear Regulatory Commission
Washington, DC 20555–0001

ABSTRACT

In NUREG-1552, "Fire Barrier Penetration Seals in Nuclear Power Plants," the U.S. Nuclear Regulatory Commission staff documented the results of its comprehensive technical assessment of penetration seals. Subsequently, the staff assessed new information for new insights. The results of the updated assessment are documented in this report. Nuclear power plants use the "defense in depth" concept of echelons of fire protection to achieve a high degree of fire safety. Fire barrier penetration seals, which are one element of the fire protection defense-in-depth concept, are designed to confine a fire to the area in which it started or to protect plant systems and components within an area from a fire outside the area. For the reasons given in this report, it is the staff's judgment that, generically, typical penetration seal deficiencies do not equate to a lack of adequate protection or result in undue risk to public health and safety. It is the staff's opinion that continued licensee attention to existing penetration seal programs and continued NRC reviews and inspections are adequate to (1) provide reasonable assurance that penetration seal problems are discovered and resolved and (2) maintain public health and safety.

TABLE OF CONTENTS

Table of Contents

_____

*Appendices A-E appear in NUREG-1552.

EXECUTIVE SUMMARY

Nuclear power plants use the "defense in depth" concept of echelons of fire protection to achieve a high degree of fire safety. The objective of this concept is to (1) prevent fires from starting; (2) rapidly detect, control, and extinguish those fires that do occur; and (3) protect structures, systems, and components important to safety so that a fire that is not promptly extinguished will not prevent the safe shutdown of the plant. The multiple layers of fire protection provided by the defense-in-depth concept offer reasonable assurance that weaknesses or deficiencies in one layer will not present an undue risk to public health and safety.

Fire barriers, which are one element of the fire protection defense-in-depth concept, accomplish their intended design function simply by remaining in place during a fire. They are important because they are the first and also the last lines of defense against a fire. That is, during the early stages of a fire, the barriers confine the fire and protect important systems and components until the fire detection and automatic fire suppression systems operate. In addition, in the event that an automatic fire protection system fails to operate or fire brigade response is delayed, the fire barriers continue to provide passive fire protection. Fire barrier penetration seals are another element of defense in depth and, like the structural fire barriers in which they are installed, are passive fire protection features. Their design function is to confine a fire to the area in which it started or to protect plant systems and components within an area from a fire outside the area. Fire barrier penetration seals are not safety related.

Between 1994 and 1996, the Office of Nuclear Reactor Regulation (NRR) staff conducted a comprehensive technical assessment of penetration seals to address reports of potential problems, to determine if there were any problems of safety significance, and to determine if NRC requirements, review guidance, and inspection procedures were adequate. The staff did not find any plant-specific problems of safety significance or any concerns with generic implications. The staff concluded that the general condition of penetration seal programs in industry was satisfactory. The staff also concluded that the information notices it had issued in 1988 and 1994, increased industry awareness of potential penetration seal problems and resulted in more comprehensive surveillance activities, maintenance practices, and corrective actions on the part of

industry. The staff concluded that these actions together with continued NRC inspections, and continued licensee attention to existing penetration seal programs, were adequate to maintain public health and safety. The staff documented its assessment in SECY-96-146, "Technical Assessment of Fire Barrier Penetration Seals in Nuclear Power Plants" (July 1, 1996), and NUREG-1552, "Fire Barrier Penetration Seals in Nuclear Power Plants" (July 1996).

The NRC staff has since continued to review potential penetration seal problems on a case-by-case basis as they are found or reported. This report supplements the NRC staff assessment of fire barrier penetration seals by reviewing additional information on seal problems reported by licensees and found during NRC inspections performed prior to as well as since the assessment documented in SECY-96-146 and NUREG-1552. In light of the new information, the staff reconsidered the operating experience reported in NUREG-1552, and considered the results of the effort, as documented in this report, for insights and appropriate opportunities for actions by the NRC and the industry.

As part of this reassessment, the staff reviewed previous NRC inspections of penetration seal programs. Between 1988 and March 1998, it conducted 153 inspections that involved installed penetration seals and penetration seal programs at 87 plants. In general, the inspectors found that the penetration seal programs were comprehensive, timely, and acceptable. In some cases, the inspectors found deficiencies and issued notices of violations. These inspections are summarized in Appendix I. In addition, the staff obtained the licensee event reports (LERs) on fire barrier penetration seals that were submitted in 1987, 1988, and 1994 through September 1998, inclusive. The staff also reviewed LERS that were submitted from 1989 through 1993 for a second time. (The staff originally documented the results of its review of these LERs in NUREG-1552.) The staff found that 9 plant sites submitted 16 LERs during 1987; 12 plant sites submitted 19 LERs during 1988; and 14 plant sites submitted 34 LERs between 1994 and September 1998. Appendix F shows the numbers of LERs and LER supplements regarding fire barrier penetration seals that were submitted by year from January 1987 through September 1998. Appendix G details the types of problems (the four major categories and subcategories) that were reported by year for the same period, and the number of times the problems occurred. Appendix H reports on each LER and LER supplement that the staff considered during this reassessment of penetration seals. This report also contains a detailed review of the status of penetration seal programs at several plants that have undertaken major corrective action programs for penetration seals.

Section III.M of Appendix R to 10 CFR Part 50 specifies that penetration seals utilize only noncombustible materials. To address questions about the NRC regulatory requirements regarding the use of these penetration seal materials, the staff reviewed the fire protection licensing basis for all nuclear plants. The staff determined which plants are required to comply with Section III.M of Appendix R to 10 CFR Part 50. The staff then conducted a detailed review of the fire protection licensing bases for those units to determine if the plants used silicone-based fire barrier penetration seal materials and, if they did, how the licensees and the staff addressed the regulatory requirement of Section III.M of Appendix R.

On the basis of everything it identified and considered, the staff judges that, overall, the issue of potential fire barrier penetration seal deficiencies is not a safety concern. For the reasons given in this report, typical penetration seal deficiencies do not necessarily equate to inadequate protection or result in undue risk to public health and safety.

On the basis of the reassessment documented here, the staff concludes that the actions it took in 1988 and 1994 to alert licensees to potential penetration seal problems increased industry awareness of such problems and resulted in more comprehensive surveillance activities, maintenance practices, and corrective actions. The staff also concludes that the general condition of penetration seal programs in industry appears to be satisfactory. The staff expects that plant-specific deficiencies may occasionally be

found during licensee surveillances and NRC inspections. However, potential penetration seal problems are understood; industry consensus fire test standards are available and are complied with; and fire test results and qualified fire-resistant seal materials and designs are available. Therefore, licensees have the means to correct problems, and continued staff oversight will ensure that corrections are made on a case-by-case basis.

In summary, it is the staff's opinion that continued licensee attention to existing penetration seal programs and continued NRC inspections are adequate (1) to ensure that penetration seal problems are discovered and resolved and (2) to maintain public health and safety.

To provide added assurance of this, during the assessment documented in this report, the staff issued Information Notice 97-70, "Potential Problems With Fire Barrier Penetration Seals," September 19, 1997, and revised the NRC fire protection core inspection module to provide more specific inspection guidance to NRC inspectors regarding fire barriers and fire barrier penetration seals. The staff will continue to assess new information regarding penetration seals for new insights and appropriate opportunities for additional actions by the staff or the industry.

During the 454th meeting of the Advisory Committee on Reactor Safeguards (ACRS), July 8-10, 1998, the staff presented the results of the assessment documented in this supplement to NUREG-1552 to the ACRS. The ACRS provided its views regarding the efforts of the NRC staff and the nuclear industry to resolve issues related to fire barrier penetration seals in a letter of July 20, 1998, from R.L. Seale, Chairman, ACRS, to Chairman Jackson. The ACRS found it clear that, overall, the NRC staff and the licensees have the issues of fire barrier penetration seals well in hand and that the efforts of the staff and the licensees have been successful in addressing the problems of the past.

1 DEFENSE-IN-DEPTH CONCEPT AND THE ROLE OF PENETRATION SEALS

1.1 Assessments of Fire Barrier Penetration Seals

Over the years, the U.S. Nuclear Regulatory Commission (NRC) staff has completed a number of assessments of fire barrier penetration seals. In 1987 and 1988, the Office of Nuclear Reactor Regulation (NRR) and regional office staff performed a comprehensive assessment of fire barrier penetration seals. Although it found no widespread problems or safety-significant generic issues, the staff alerted industry to potential problems by means of a series of information notices. Later, in 1993, NRR staff reassessed the fire protection program for nuclear reactors. In its "Report on the Reassessment of the NRC Fire Protection Program" (February 27, 1993), the staff concluded that licensees were complying with regulatory requirements and that there were no major or recurring issues with penetration seals. In 1995, the Office for the Analysis and Evaluation of Operational Data (AEOD) reviewed fire barrier penetration seals and reached many of the same conclusions that NRR had reached. Finally, between 1994 and 1996, NRR staff conducted a comprehensive technical assessment of penetration seals to address reports of potential problems, to determine if there were any problems of safety significance, and to determine if NRC requirements, review guidance, and inspection procedures are adequate. The staff did not find any safety-significant plant-specific problems or concerns with generic implications. The staff concluded that the general condition of penetration seal programs in the nuclear industry was satisfactory. The staff also concluded that the information notices it had issued in 1988 and 1994 increased industry awareness of potential penetration seal problems and resulted in more comprehensive surveillance activities, maintenance practices, and corrective actions. Moreover, the staff concluded that these staff actions, together with continued licensee attention to existing penetration seal programs and continued NRC inspections, were adequate to maintain public health and safety. The staff documented its assessment in SECY-96-146, "Technical Assessment of Fire Barrier Penetration Seals in Nuclear Power Plants" (July 1, 1996), and NUREG-1552, "Fire Barrier Penetration Seals in Nuclear Power Plants" (July 1996).

Notwithstanding these findings, the NRC staff reviews potential problems on a case-by-case basis as they are found or reported. Therefore, the NRC staff updated its assessment of fire barrier penetration seals by assessing information on seal problems reported by licensees and found during NRC inspections since the assessment documented in SECY-96-146 and NUREG-1552. The staff reconsidered the operating experience reported in NUREG-1552 in light of the new information, and also considered the results of this effort, which is documented herein, for insights and appropriate opportunities for actions by the NRC and the industry.

1.2 The Role of Penetration Seals in the Defense-in-Depth Concept

Nuclear power plants use the "defense in depth" concept of echelons of fire protection to achieve a high degree of fire safety. The objective of the concept is to (1) prevent fires from starting; (2) promptly detect, control, and extinguish those fires that do occur; and (3) protect structures, systems, and components important to safety so that a fire that is not promptly extinguished will not prevent the safe shutdown of the plant. The several layers of fire protection produced by the defense-in-depth concept offer reasonable assurance that weaknesses or deficiencies in one layer will not present an undue risk to public health and safety. To achieve defense in depth, each operating reactor maintains an NRC-approved fire protection program. The licensees have designed the fire protection programs by analyses that (1) considered potential fire hazards, (2) determined the effects of fires in the plant on the ability to safely shut down the reactor or on the ability to minimize and control the release of radioactivity to the environment, and (3) specified measures for fire prevention, fire confinement, fire detection, automatic and manual fire suppression, and post-fire safe-shutdown capability.

Nuclear power plants are divided into separate areas by structural fire barriers such as concrete floors, walls, and ceilings. The fire protection function of these barriers is to prevent a fire that starts in one plant area from spreading to another area. A barrier's fire-resistance rating, which is a measure of the extent to which the barrier resists the effects of fire, is determined by exposing a mockup of the barrier to an intense test fire for a designated period. Nuclear power plant fire barriers typically have a fire-resistance rating of 1, 2, or 3 hours. Openings are needed in structural fire barriers to allow such items as cable trays, conduits, pipes, and ventilation ducts to pass from one plant area to another. To maintain the fire protection function of the structural fire barriers, the openings and the gaps and annular spaces around the penetrating items (penetrations) should be sealed in a configuration that offers the same fire resistance as that of the barrier in which they are installed. The average number of fire barrier penetration seals per nuclear power plant unit is about 3000 and a single unit can have up to 10,000 seals.

Fire barriers, which are one element of the fire protection defense-in-depth concept, accomplish their intended design function simply by remaining in place during a fire. They are important because they may serve as the first and also the last lines of defense against a fire. That is, during the early stages of a fire, the barriers confine the fire and protect important systems and components until the fire detection and automatic fire suppression systems operate. In addition, in the event that an automatic fire protection system fails to operate or fire brigade response is delayed, the fire barriers continue to provide passive fire protection. Fire barrier penetration seals are another element of defense in depth and, like the structural fire barriers in which they are installed, are passive fire protection features. Their design function is to confine a fire to the area in which it started or to protect plant systems and components within an area from a fire outside the area.

To gain reasonable assurance that a penetration seal will have the required fire-resistance capability or fire rating, a penetration seal test assembly is subjected to a fire endurance test. The test methods involve the furnace-fire exposure of a full-scale penetration seal test specimen that is representative of the construction for which a fire-resistance rating is desired. The heat input to the test furnace is controlled so that the average temperature in the furnace follows the time-temperature curve specified in the test standard. In the United States, the standards for testing penetration seals use the time-temperature curve defined in American Society for Testing and Materials (ASTM) E-119, "Standard Test Methods for Fire Tests of Building Construction

and Materials[1]." This time-temperature curve, which is used to determine the fire resistance of all types of building fire barriers, represents a severe fire exposure. (It is important to note that fire tests are not intended to model any specific room fire or the conditions under which the seals will be exposed during a fire, but rather to provide a specific standard fire exposure against which similar fire rated assemblies can be evaluated.)

The fire protection effectiveness of structural fire barriers is largely dependent on their inherent fire resistance, details of construction, and protection of penetrations. Some fire barriers (both structural barriers and penetration seals) are more important to the fire protection defense-in-depth concept than others. The importance of specific fire barriers depends on many factors, such as the importance of the plant systems and components in the fire area (and adjacent areas); the types, amounts, configurations, and locations of combustible materials and fire hazards, if any, in the areas; the potential for fire growth in the areas; the fire protection features installed in the areas; and the accessibility of the areas to the plant fire brigade. The importance of specific penetration seals depends on these factors and on such other factors as their size, their location or position in the fire barrier, and the number and sizes of the other seals in the barrier.

In order of overall importance to fire protection defense in depth, structural fire barriers, being necessary for the structural integrity of a building or fire area, are generally considered to be more important than fire barrier penetration seals. Qualified fire protection engineers determine the importance of individual fire barriers by analyzing fire hazards and the locations of safe shutdown and safety-related systems and components.

Although a detailed discussion of such analyses is beyond the scope of this paper, the following discussion illustrates this approach.

Consider, for purposes of a worst-case analysis, that a structural fire barrier fails during exposure to a fire. In this event, the adjoining fire area and its contents would be exposed to the same fire and would, themselves, become involved in the fire in a short

---

[1]Representative points on the curve that determine its character are: 1000 °F at 5 minutes, 1550 °F at 30 minutes, 1700 °F at 1 hour, 1850 °F at 2 hours, and 1938 °F at 3 hours.

period of time. (Because of the substantial construction of structural fire barriers in nuclear power plants and fire protection defense in depth, the staff does not consider this a credible nuclear power plant fire scenario.) Similarly, catastrophic failure of a penetration seal could expose the adjacent fire area to the fire. However, since the penetration seal is not necessary for structural integrity, its failure is not as significant a fire threat as the failure of a structural fire barrier would be. In addition, in most cases, a seal failure would initially create a localized hot spot in the adjacent fire area in the area of the seal. If there are no combustible materials in the adjacent fire area in the vicinity of the failed seal (for example, if the penetration seal surrounds a pipe), smoke and hot gases will migrate into the adjacent area, but the spread of fire into the area will be limited. If there are combustible materials in the vicinity of the failed seal (for example, if the penetration seal surrounds a loaded cable tray that passes from one fire area to another), the fire could spread into the adjacent area more readily. In this instance, a more detailed fire hazards analysis is needed to assess the potentially adverse effects of the fire spread. Regardless, such a fire scenario is less threatening than the failure of a structural fire barrier.

2 REVIEW OF REACTOR OPERATING EXPERIENCE

2.1 Licensee Event Reports

In NUREG-1552, the staff reported that in 1994 the licensee event report (LER) database maintained by Oak Ridge National Laboratory contained about 58,000 LERs and that 318 (about 0.5 percent) of them, involved fire barrier penetrations. (For this discussion, "LERs" also includes LER supplements.)

In NUREG-1552 the staff documented the results of its review of the LERs submitted between 1989 and 1993, inclusive. The staff found that licensees for about 20 plant sites had submitted 141 LERs regarding fire barrier penetration seals. In support of the reassessment documented here, the staff obtained the LERs regarding fire barrier penetration seals that were submitted in 1987 and 1988, and 1994 through September 1998, inclusive. The staff found that 9 plant sites submitted 16 LERs during 1987; 12 plant sites submitted 19 LERs during 1988; and 14 plant sites submitted 34 LERs between 1994 and September 1998.

Overall, the staff found that the technical problems with penetration seals that were reported between 1987 and September 1998, inclusive, could be classified into four major categories. In descending order of the number of reported occurrences, these were

(1) seal not installed or breached (58 occurrences),

(2) seal not properly installed (63 occurrences),

(3) inadequate documentation (19 occurrences), and

(4) seal degraded or damaged (17 occurrences).

Appendix F shows the numbers of LERs regarding fire barrier penetration seals that were submitted by year from January 1987 through September 1998. Appendix G details the types of problems (the four major categories and subcategories) that were reported by year for the same period, and the number of times the problems occurred. Appendix H reports on each LER that the staff considered during this reassessment of penetration seals. (The total number of LERs for 1989 through 1993 differs from the number reported in NUREG-1552 because the staff removed from consideration reports that were not related to technical problems, e.g., missed surveillances. Note also that some licensees do not consider that penetration seal deficiencies are conditions that put a plant outside its design basis and, therefore, do not report such deficiencies in LERs.)

As part of this reassessment, the staff reviewed the LERs submitted during 1987 and 1988 and those submitted from 1994 through September 1998. The staff also reconsidered the LERs that were submitted from 1989 through 1993. On the basis of its reviews, the staff made the following observations:

(1) The types of problems that were reported during 1987 and 1988 and from 1994 through 1998, were consistent with the types of problems reported in the LERs submitted from 1989 through 1993. The staff did not uncover new types of problems.

(2) It appears that the types of problems and deficiencies that have been found (e.g., voids, cracks, inadequate documentation) have involved each type of seal used by industry (e.g., grout, silicone foam, and silicone elastomer).

(3) Overall, the number of LERs submitted each year has decreased from a high of 23 in 1989 to 8 in 1998 (through September).

(4) The number of occurrences of penetration seal deficiencies has decreased from a high of 25 in 1989 to 7 in 1998 (through September).

(5) After its first comprehensive technical assessment of fire barrier penetration seals, the NRC staff issued Information Notices (INs) 88-04; 88-04, Supplement 1; and 88-56 to alert industry to potential seal problems. In response to these INs, there was significant industry scrutiny of installed penetration seals and penetration seal programs. On the basis of its best-effort search of LERs and NRC inspection reports (see Section 2.2, below), the staff found that the licensees for at least 45 plants have conducted enhanced[2] 100-percent penetration seal inspections in response to the INs. (See Appendix J for a complete list of references.)

(6) Most of the licensees that have conducted 100 percent seal inspection programs found seal deficiencies. The findings ranged from negligible to widespread problems involving each of the four categories of problems. These licensees strengthened their programs to reduce the likelihood of recurrence.

(7) Many of the deficiencies concerning failure to install seals, improper seal installation, and inadequate documentation existed since the plant was built. However, these types of problems can occur at any time during the life of the plant. For example, during plant outages, temporary and permanent modifications that involve routing cables are commonplace. Such modifications require breaching existing penetration seals or making new penetrations. Plant procedures specify that the breached seals be restored and that new penetrations be sealed with properly designed and tested penetration seal

---

[2]For purposes of this discussion, an enhanced program is one that exceeds the requirements of the licensee's routine surveillance program. For example, the licensee may have compared test documentation to installed seal configurations or removed damming boards to verify the thickness of the installed seals.

assemblies. Sometimes this is not done and the discrepancies are not found until a subsequent penetration seal surveillance.

(8) In some cases, licensees conservatively reported such superficial problems as surface imperfections and small cracks, splits, and gaps, which would not have precluded the seals from performing their intended fire protection design function.

(9) Licensees appear to understand potential problems with and corrective actions for fire barrier penetration seals.

(10) Plant age does not appear to be a critical attribute as to whether or not a plant is prone to seal problems. Of the 45 plants known to have completed 100-percent seal inspection programs, about half operated before January 1, 1979 (and are covered by the regulations in Appendix R), and half began operations later and are not covered by the regulations in Appendix R.

(11) Overall, the safety significance and risk significance of the reported deficiencies were low. The potential safety significance of the reported problems is discussed in Section 3. The risk significance is discussed in Section 4.

Of the LERs submitted since the staff issued NUREG-1552, two indicated widespread plant-specific deficiencies. The first involved Washington Nuclear Project 2 (WNP2) and the second involved Maine Yankee. The staff was aware of the deficiencies at WNP2 through previous NRC inspections and it documented these deficiencies and the licensee's corrective actions in Section 5.5.5 of NUREG-1552. The staff's assessment of the Maine Yankee report is in Section 6.6 of this report.

2.2 NRC Inspections

As part of this reassessment, the staff conducted a best-effort search for NRC inspections of penetration seal programs. The staff found that between 1988 and March 1998, it conducted 153 inspections that involved installed penetration seals and penetration seal programs at 87 plants. Of these, 42 (48 percent) were Appendix R plants (operating prior to January 1, 1979). The inspectors reviewed the adequacy of penetration seal installations, qualification, and surveillances. They also followed up on issues reported in LERs and weaknesses noted during previous NRC inspections. In some cases, the inspectors reviewed the 100-percent penetration seal reevaluation programs performed by the licensees. In other cases, the inspectors walked down the seal installations to assess their adequacy. In general, the inspectors found that the penetration seal programs were comprehensive, timely, and acceptable. In some cases, the inspectors found deficiencies and issued notices of violations. Each of these inspections is summarized in Appendix I.

On the basis of its review of the NRC inspection findings, the staff made the following observations:

(1)   The types of problems found during inspections were consistent with the types of problems reported in LERs. The staff did not identify new types of problems during its inspections.

(2)   The inspection reports, like the LERs, revealed that licensees occasionally find plant-specific deficiencies.

(3)   For the most part, the licensees maintained satisfactory fire barrier penetration seal programs.

(4)   Licensees understand potential fire barrier penetration seal problems, have the means to correct problems, and have taken appropriate and timely actions to correct penetration seal deficiencies.

(5)   The NRC inspection reports did not reveal widespread or potentially generic problems of safety significance.

As noted in NUREG-1552, the NRC's routine fire protection inspection procedures are contained in the NRC Inspection Manual in Inspection Procedure 64704, "Fire Protection Program" (March 18, 1994). This procedure directs the inspectors to visually inspect the fire barriers associated with two plant fire areas and ensure that the electrical and mechanical penetration seals are functional. However, the procedure did not give specific guidance for inspecting the seals or establishing their functionality. The lack of specific inspection guidance was viewed as a potential weakness in the NRC reactor fire protection program. Therefore, the staff revised Procedure 64704 in September 1997, to add guidance for inspecting penetration seals as a part of its routine fire protection inspections.

In NUREG-1552, the staff also reported that it was preparing the new fire protection functional inspection (FPFI) program that it had described in SECY-95-034, "Status of the Recommendations Resulting from the Reassessment of the NRC Fire Protection Program." Since it issued NUREG-1552, the staff has drafted the FPFI procedures and guidelines and has started the pilot FPFI program. The FPFI procedures and guidance contain detailed guidance for inspecting fire barrier penetration seals and seal programs. These procedures and guidelines are being used during the FPFIs and are available for NRC inspectors and licensees to use on an as-needed basis independent of an FPFI.

2.3   Fire Experience

The staff reviewed the fire event databases compiled by Sandia National Laboratories, which contained data from 1965 thorough 1985, and the Electric Power Research Institute, which contained data from 1965 through 1988. The staff found no reports of nuclear power plant fires that challenged the ability of fire rated structural barriers or fire rated penetration seals to confine a fire in accordance with their fire protection design function. The staff also reviewed the LER database discussed in Section 2.1, which contains data from 1980 to the present, and again, found no reports of nuclear power plant fires that caused the failure of a fire rated structural barrier or a fire rated penetration seal. In addition, since the staff issued NUREG-1552, AEOD issued a special study titled "Fire Events—Feedback of U.S. Operating Experience" (June 1997), which covers operating experience from 1965 through 1994. This AEOD study does not contain fire events that challenged either fire-rated structural barriers or fire-rated penetration seals.

It has been suggested that the March 22, 1975, fire at the Browns Ferry Nuclear Plant propagated through a fire-rated penetration seal and, therefore, there is industry experience that a fire challenged such a seal. The staff does not agree. As reported in NUREG-0050, "Recommendations Related to Browns Ferry Fire" (February 1976), "the seal that caught fire differed from the [fire] seal as designed and tested." For example, the installed seal in which the fire started used flexible polyurethane foam rather than the spray polyurethane foam specified in the plant's original design criteria. In addition, the installed seal did not have the fire-retardant coating specified in the design criteria. Furthermore, the report stated that "a properly made fire stop of the

Browns Ferry design (with Flammastic and without flexible foam) would probably not have initiated the fire" and "even if a fire had started, a fire stop made in accordance with the original design may well have prevented its spread outside of the room where it started."

2.4 Summary of Operating Experience

The LERs and NRC inspection reports show that many plants have performed 100-percent penetration seal inspections and corrective action programs since 1987. The staff found no evidence of generic problems of safety significance with penetration seal materials or safety-significant failures of penetration seals. On the basis of its review, the staff concluded that the licensees have been effective in finding penetration seal deficiencies and have taken timely and appropriate actions to correct identified discrepancies. In view of the large number of penetration seals installed in nuclear power plants, the staff expects that plant-specific deficiencies may occasionally be found during licensee surveillances and NRC inspections. However, the LERs and NRC inspection findings show that licensees understand the potential fire barrier penetration seal problems and that fire test results and qualified fire-resistant seal materials and designs are available. Therefore, licensees have the means to correct problems. Appendix J lists plants that, on the basis of docketed information, are known to have performed 100-percent penetration seal inspection programs that exceeded the specifications of the licensees' normal fire barrier surveillance programs. Appendix K lists the docketed references (LERs and NRC inspection reports), by plant, that the staff considered in this reassessment of fire barrier penetration seals.

3 SAFETY SIGNIFICANCE

3.1 Fire Protection Program

The basic fire protection regulation for commercial nuclear power plants is Title 10 of the *U.S. Code of Federal Regulations*, Part 50, Section 50.48, "Fire protection." Section 50.48(a) states that each operating nuclear power plant must have a fire protection plan that satisfies General Design Criterion (GDC) 3 of Appendix A to 10 CFR Part 50, "Fire protection," and notes that fire protection guidance for nuclear power plants is contained in

Branch Technical Position (BTP) Auxiliary Power Conversion Systems Branch (APCSB) 9.5-1, "Guidelines for Fire Protection for Nuclear Power Plants;" and Appendix A to BTP APCSB 9.5-1, "Guidelines for Fire Protection for Nuclear Power Plants Docketed Prior to July 1, 1976." These two NRC documents specify preferred methods for fire protection program design. In addition, Section 50.48(b) states that Appendix R to 10 CFR Part 50, "Fire Protection Program for Nuclear Power Facilities Operating Prior to January 1, 1979," establishes fire protection features required to satisfy GDC 3 with respect to certain generic issues for nuclear power plants licensed to operate before January 1, 1979. Fire protection programs that meet the criteria of either BTP APCSB 9.5-1 or Appendix A to BTP APCSB 9.5-1 and the applicable sections of Appendix R satisfy 10 CFR 50.48 and GDC 3. NUREG-0800, "Standard Review Plan," (SRP) Section 9.5-1, "Fire Protection Program," incorporates the guidance of BTP APCSB 9.5-1 and Appendix A to BTP APCSB 9.5-1 and the criteria of Appendix R. Therefore, fire protection programs that meet the guidelines of SRP Section 9.5-1 also satisfy 10 CFR 50.48 and GDC 3.

The objective of the fire protection program required by 10 CFR 50.48 is to minimize both the probability and consequences of fires. As discussed in Section 1, the licensees use the concept of defense in depth to achieve a high degree of fire safety. The licensees determine the adequacy of fire protection for plant safety systems and areas by analyzing the effects of postulated fires. In general, the primary means of fire protection consists of fire barriers and fixed automatic fire detection and suppression systems. In addition, manual fire fighting capability is provided throughout the plant to limit the extent of fire damage. The plant fire hazards analysis addresses the following variables and attributes:

(1) the NRC fire protection requirements and guidance that apply;

(2) amounts, types, configurations, and locations of cable insulation and other combustible materials;

(3) fire loading and calculated fire severities;

(4) in situ fire hazards;

(5) automatic fire detection and suppression capability;

(6) layout and configurations of safety trains;

(7) reliance on and qualifications of fire barriers, including fire test results, the quality of the materials and system, and the quality of the installation;

(8) fire area construction (walls, floor, ceiling, dimensions, volume, ventilation, and congestion);

(9) location and type of manual fire fighting equipment and accessibility for manual fire fighting;

(10) potential disabling effects of fire suppression systems on shutdown capability;

(11) availability of oxygen to support combustion (for example, inerted containment); and

(12) post-fire safe-shutdown capability, including alternative or dedicated shutdown capability.

During its reviews and inspections of the licensees' fire protection programs, the staff ensured that each licensee had provided an adequate level of fire protection.

3.2 Safety Significance Ranking of Penetration Seal Deficiencies

In general, the potential safety significance of a deficient fire barrier penetration seal depends on such factors as the nature and extent of the deficiency; the importance of the plant systems and components in the fire area (and adjacent areas); the amounts, types, configurations, and locations of any combustible materials and fire hazards in the areas; the potential for fire growth in the areas; the fire protection features installed in the areas; and the accessibility of the areas to the plant fire brigade. The actual safety significance and the importance of a specific seal depends on these factors and on such other factors as its size, its location or position in the fire barrier, and the number and sizes of the other seals in the barrier.

Appendix G summarizes the types of penetration seal problems and deficiencies that were reported in LERs, by year, from 1987 through September 1998, inclusive. It is the staff's judgment that, in general, the four categories of deficiencies presented in Section 2.1 of this report and in Appendix G can be ranked from highest potential safety significance to

lowest as follows: (1) seal not installed or breached, (2) seal not properly installed, (3) seal degraded or damaged, and (4) inadequate documentation.

3.3 Generic Assessment of Safety Significance

For purposes of the following discussion, the safety significance of a fire barrier penetration seal can be thought of as being the role the seal plays in preventing a fire from spreading from the fire area of origin to an adjacent fire area. In the *Federal Register* notice that issued the proposed Appendix R to 10 CFR Part 50,[3] the staff stated that the "phenomenon of fire is believed to be sufficiently well understood to permit evaluation of existing and potential fire hazards and probable extent of damage should a fire occur. Such evaluations are useful in assessing the possible consequences of fire in a given area." In this regard, a generic assessment is instructive for understanding the safety significance of fire barrier penetration seals.

As discussed in Section 1, licensees rely on a defense-in-depth concept that incorporates several fire safety measures. In sum, automatic fire detection and suppression systems are provided in most areas that have safe-shutdown equipment. Trained fire brigades are required to be on duty 24 hours a day at all plants. All areas that have safe-shutdown equipment contain manual fire suppression features. Fuels that can feed a fire and ignition sources to start a fire are controlled. Taken together, these factors represent an adequate means of fire protection at the plants and ensure that operations can be conducted without an undue risk to the health and safety of the public. In general, every echelon of fire protection defense in depth would have to either fail or be significantly compromised for a fire to breach a fire barrier penetration seal and adversely affect the safe-shutdown capability or cause other operational problems. Specifically, the following would have to occur:

(1) Despite the plant fire prevention program, a fire would have to occur.

(2) The fire would have to go undetected. That is,

[3]U.S. NRC, "Fire Protection Program for Nuclear Power Plants Operating Prior to January 1, 1979," *Federal Register*, Vol. 45, No. 105, May 29, 1980, pp. 36082–36090.

the automatic fire detection and alarm system would have to fail. In addition, plant personnel would have to fail to discover the fire.

(3) The fire would have to grow beyond the incipient stage, spread, and become large. This means that the fire area would have to contain transient and *in situ* combustible materials of sufficient types, amounts, and configurations to support fire growth and spread.

(4) The automatic fire suppression system (if there is one) would not operate and control the fire, or if it operated, it would fail to control the fire.

(5) Manual fire suppression activities would not be employed to control and suppress the fire.

(6) The fire must expose the safe-shutdown components located in the originating fire area and cause fire damage that renders the components nonfunctional. For this to happen, the fire must either start near the components or it must spread close enough to the components so that the components are damaged by direct flame impingement or radiative heat transfer. Alternately, the fire's products must adversely affect the safe-shutdown components located in the fire area. For example, hot gases from the fire would rise to the ceiling and form a hot gas layer. Safe-shutdown components (e.g., cables) located near the ceiling and within the hot gas layer could be damaged by the convected heat even if they are located away from the burning area.

(7) The fire must also spread to a penetration seal installed in a structural fire barrier that separates the fire area of origin from an adjacent fire area with the other train of redundant safe-shutdown components.

(8) The uncontrolled fire must burn through the fire-resistant penetration seal assembly (which in some cases, could take more than 3 hours).

(9) After the fire burns through the penetration seal, it must continue to burn and spread from the penetration to the redundant safe-shutdown components located in the adjacent fire area, where it must cause sufficient fire damage to

the components to affect their ability to function. That is, the scenario described under items 1 through 6 would also have to occur in the second fire area.

As discussed in Section 1, fire barrier penetration seals are passive fire protection features that accomplish their intended fire protection function by their very presence. Penetration seals are important features because they help confine a fire to its area of origin. There can be no question that when properly designed and installed, the various types of penetration seals currently installed in nuclear power plants will provide fire resistance equivalent to the barriers in which they are installed and will perform their intended fire protection function by confining a fire to the area of origin. The types of penetration seal deficiencies described in Section 2 and in Appendix G can reduce the fire-resistance capabilities of penetration seals. Nevertheless, it is the staff's opinion that, in general, the relative safety significance of such deficiencies is low for the following reasons: in most cases, the deficiencies may reduce the fire resistance of the seal, but they do not render it useless; the defense-in-depth concept ensures that multiple safety measures are incorporated; automatic fire detection and suppression systems are provided in areas that have safe-shutdown systems and components; trained fire brigades are required to be on duty 24 hours a day at all plants; and transient and *in situ* fuels and fire hazards that can feed a fire, and ignition sources that can start a fire, are controlled. Therefore, it is unlikely that a fire significant enough to challenge a fire barrier penetration seal will occur. How these factors affect the various types of penetration seal deficiencies is discussed in Sections 3.3.1 and 3.3.2, below.

3.3.1 Improperly Installed or Degraded Seals and Inadequate Documentation

As discussed in Section 1, the fire endurance tests maximize fire severity by subjecting the penetration seal to a fire of rapidly rising temperature in a relatively small and confined space. In the event of an actual fire at a nuclear power plant, the fire resistance required of a penetration seal depends on the expected severity of the fire to which it may be exposed. With few exceptions, nuclear plant fire loads are not great enough to produce a fire approaching the severity of a test fire (time and temperature). It is expected that the temperature of most actual fires at nuclear power plants would rise slower than the temperature of the standard test fire. Most plant areas have controls on ignition sources; these controls help reduce the occurrences of fires. Most plant areas are equipped with other passive and active fire protection features, and many are continuously or regularly occupied by plant operators, security staff, and other personnel, all of whom contribute to early fire detection and suppression activities. For example, plant fire detection systems give reasonable assurance that a fire will be detected in its incipient stage and before there is any significant propagation of flame; or rise in temperature. The detection system would send an alarm to the continuously manned control room, and the control room operators would dispatch the plant fire brigade. The fire brigade would then extinguish the fire.

In a plant area that is protected by an automatic fire suppression system, should the fire develop beyond the incipient stage before the fire brigade responds, the system would actuate and either control or extinguish the fire. Therefore, there is reasonable assurance that a fire will not challenge a fire barrier penetration seal.

In addition, in large open spaces, such as exist in many nuclear plant fire areas, a fully developed fire may occur in one part of the area (e.g., in concentrations of cables), but it is not probable that the entire volume (fire area) would be engulfed in flames (flashover) before an automatic fire suppression system actuated or manual fire suppression activities were employed. Unless a fire reaches the fully developed stage, it is not likely to present a credible challenge to any nuclear power plant penetration seal. Moreover, even in cases in which the fire barrier penetration seals are degraded or deficient, they will offer some measure of fire protection. Some of the reported deficiencies could have reduced the fire-resistance rating of seals under test conditions and the fire protection effectiveness of in-plant seals (e.g., inadequate seal thickness). However, other deficiencies (splits, shrinkage, inadequate documentation) may have little or no effect on seal performance.

3.3.2 Unsealed and Breached Penetrations

For the cases discussed in Section 3.3.1, the installed penetration seals are degraded or deficient, but will provide some measure of fire protection. Intuitively, conditions involving missing and breached seals

involve potentially higher safety significance, because this measure of protection is missing altogether and the fire may have a direct path to spread from one fire area to another.

It is important to note that there is *no* regulatory requirement that fire-rated seals be installed in all penetrations through fire barriers that form fire area boundaries or that seals have either (1) the same fire-resistance rating as the structural fire barrier in which they are installed or (2) a 3-hour fire resistance rating. In Generic Letter (GL) 86-10, "Implementation of Fire Protection Requirements" (April 24, 1986), the staff presented guidance for satisfying NRC regulatory requirements for fire protection. In Enclosure 1 to GL 86-10, the staff interpreted Appendix R requirements. Interpretation 4, "Fire Area Boundaries," stated, in part,

> The term "fire area" as used in Appendix R means an area sufficiently bounded to withstand the [fire] hazards associated with the area and, as necessary, to protect important equipment within the area from a fire outside the area. In order to meet the regulation, fire area boundaries need not be completely sealed floor-to-ceiling, wall-to-wall boundaries. However, all unsealed openings should be identified and considered [in] evaluating the effectiveness of the overall barrier. Where fire area boundaries are not wall-to-wall, floor-to-ceiling boundaries with all penetrations sealed to the fire rating required of the boundaries, licensees must perform an evaluation to assess the adequacy of fire boundaries in their plants to determine if the boundaries will withstand all [fire] hazards associated with the area.

This regulatory position established that certain penetration seals need not have the same fire rating as the barrier in which they are installed and, indeed, that certain fire barrier penetrations may not need to be sealed at all. Licensees evaluate such seals on a case-by-case basis. The engineering evaluations performed to assess the effectiveness of the penetration seals are based on the expected fire-resistive performance of the seal and on the fire hazards and fire protection features in the fire area. Nevertheless, on the basis of its experience, the staff believes that most licensees install 3-hour fire-rated penetration seals in fire area boundaries.

It should be noted that with up to 10,000 fire barrier penetration seals per nuclear unit, the instances of unsealed penetrations and breached penetration seals that have been reported are rare. Open penetrations are more safety significant than degraded penetration seals. However, even in cases of missing or breached seals, most of the considerations discussed in Section 3.3.1 still apply. That is, the defense-in-depth concept ensures that multiple safety measures are incorporated; automatic fire detection and suppression systems are provided in areas that have safe-shutdown systems and components; trained fire brigades are required to be on duty 24 hours a day at all plants; and transient and *in situ* fuels and fire hazards that can feed a fire and ignition sources that can start a fire are controlled. To spread through an open penetration, the fire would have to be large and uncontrolled. In this case, a localized hot spot would occur in the adjacent fire area in the area of the seal. If there are no combustible materials in the adjacent fire area in the vicinity of the open penetration (for example, if the penetration seal encloses a pipe), smoke and hot gases will move into the adjacent area, but the spread of fire into the area would be limited. Conversely, if there are combustible materials in the vicinity of the failed seal (for example, if the penetration seal encloses a loaded cable tray that passes from one fire area to another), the fire could spread into the adjacent area more readily. However, in the event a fire spreads through an unsealed penetration, the fire threat to the adjoining fire area should be readily mitigated by the plant fire brigade.

As an example, consider the following. On March 22, 1975, the Browns Ferry Nuclear Power Plant had the worst fire ever to occur in a commercial nuclear power plant operating in the United States. As reported in NUREG-0050, the fire spread along cable trays from the cable spreading room, through a cable penetration, and into the reactor building. The fire burned cables in cable trays for almost 7 hours. During that time, portable extinguishers were used intermittently to no effect. After almost 7 hours, the decision was made to fight the fire with water. Two men using a fire hose extinguished the fire within 15 minutes. This experience demonstrated that a significant and challenging nuclear power plant fire could be readily extinguished if appropriate and timely fire fighting efforts are employed. Since the fire at Browns Ferry, licensees have made significant improvements in fire brigade training and fire fighting capabilities. The staff believes that if timely and appropriate action is initiated, a fire at an open penetration will not create any significant problems. Therefore, on the aforementioned bases, although the staff considers an open penetration to be more significant than a degraded seal, it believes that the relative safety significance of missing and breached seals, although potentially higher than the other common types of seal deficiencies, is low.

3.4 Seal-Specific Assessment of Safety Significance

For the reasons discussed above, in general, the safety significance of deficient fire barrier penetration seals is low. However, the actual safety significance of specific deficiencies in fire barrier penetration seals depends on many factors and variables. These include the importance of the plant systems and components in the fire area (and adjacent areas); the types, amounts, configurations, and locations of any combustible materials and fire hazards in the areas; the potential for fire growth in the areas; the fire protection features installed in the areas; the accessibility of the areas to the plant fire brigade; the type, size, and location of the penetration seal; the nature and extent of the seal deficiencies; and the overall effectiveness of the defense-in-depth process.

Clearly, certain fire areas present a more credible challenge to deficient fire barrier penetration seals than others. For example, it is likely that a fire involving a turbine generator lubricating oil system would present a significant fire exposure to the fire barrier penetration seals installed in the fire wall that separates the turbine building from the auxiliary building. If the seals are properly designed and installed and the other components of the fire protection program (e.g., fire brigade) are effective, they are likely to withstand the challenge and prevent the fire from spreading from the turbine building into the auxiliary building. However, if the seals are deficient, it is conceivable that they could fail under the fire exposure and allow the fire to spread into the auxiliary building. Again, the actual adverse consequences of this situation would depend on such factors as the location of the burnthrough into the auxiliary building and the location of combustibles and important plant equipment in the vicinity of the burnthrough. The significance of such a scenario could be compounded by the fact that the fire wall in the turbine building could be common to several auxiliary building fire areas. Therefore, if the penetration seals were to fail, a single fire could adversely impact several plant components and systems.

On the other hand, a fire involving a charging pump motor is not likely to present nearly as significant a challenge to fire barrier penetrations installed in the pump cubicle walls. In this case, even if the seals are deficient, the fire is not likely to have an adverse effect on plant safety systems located outside of the pump cubicle.

4 RISK SIGNIFICANCE

The calculated core-damage frequency (CDF) from fires, and the contribution of fire risk to a plant's total CDF, is a plant-specific determination that is dependent on the plant configuration and the methodology and assumptions that are used for the analysis. The application of the calculated CDF to assess the fire risk of one plant against the fire risk at another plant is inappropriate.

The postulated fire scenarios that are the major contributors to core damage for most plants are those in which the redundant divisions of post-fire safe-shutdown components and systems are located in the same fire area. In these scenarios, fire barrier penetration seals are not considered (not modeled) in the assessment, because the factors mentioned earlier have a greater effect on CDF.

Scenarios involving the spread of fire from one plant fire area to another and evolving to core damage are of low frequency. This is a result of several defense-in-depth measures, such as administrative controls on combustible materials and "hot" work, automatic fire detection, automatic fire suppression, and intervention by the plant fire brigade. On the basis of its reviews of fire risk assessments completed thus far, penetration seals have not been relied upon for the prevention of core damage. It is the staff's judgment that failure of a plant's barrier penetration seals would not significantly alter the overall contribution of fire risk to the plant's total calculated CDF.

5 COMPENSATORY MEASURES

The use of fire watches in instances of degraded or inoperable fire barriers is an integral part of NRC-approved fire protection programs. In general, these approved compensatory measures specify the establishment of a continuous "fire watch" or an hourly fire watch patrol where automatic detection systems protect the affected components. Fire watches are personnel trained by the licensees to inspect for the control of ignition sources, fire hazards, and combustible materials; to look for signs of incipient fires; to provide prompt notification of fire hazards and fires; and, in some cases, to take appropriate actions to begin fire suppression activities. Generally, therefore, by providing additional fire prevention activities through enhanced capabilities to find fire hazards and, in the case of a fire, through augmented suppression activities before a penetration seal's ability to endure a fire is challenged, fire watches compensate for degraded fire barrier penetration seals. The licensees that reported fire barrier penetration seal deficiencies established fire watches in accordance with their technical specifications or license conditions as a compensatory measure.

6 PLANT-SPECIFIC EXPERIENCE WITH FIRE BARRIER PENETRATION SEALS

The staff reviewed in detail the status of penetration seal programs at several plants that have undertaken major penetration seal corrective action programs.

6.1 Vermont Yankee Nuclear Power Station

On March 19, 1992, during an inspection of fire barrier penetration seals at Vermont Yankee Nuclear Power Station, the licensee found a penetration containing unapproved material. The next day, another penetration seal was found to be degraded. The licensee took compensatory measures and began an investigation into the cause of the degradation. Later, while implementing corrective actions in December 1992, the licensee found more problems. It performed additional seal inspections and found that the seal discrepancies were more widespread than was originally believed. On January 15, 1993, the licensee issued Licensee Event Report (LER) 93-001. The licensee declared 57 penetration seals inoperable and established a task force to inspect all fire barrier penetration seals. Ultimately, the licensee repaired more than 900 (64 percent) of the 1400 fire barrier penetrations installed at Vermont Yankee and upgraded almost 300 penetrations (21 percent). The licensee attributed most of the as-found unacceptable penetrations to inadequate design or to inadequate installations made by a contractor between 1979 and 1980. (That contractor is no longer in business.) The licensee attributed the failure to identify these issues to inadequate surveillance procedures. The licensee completed the repairs to affected barriers and the required surveillances in May 1993. In subsequent years (1994–1997), routine fire barrier surveillances discovered five degraded penetration seals. These events were described in LERs 94-018, 94-018-01, 95-004, 96-026, and 96-026-01. In 1998, the licensee reported seal problems in LERs 98-001, 98-001-01, 98-008, 98-008-01, 98-014, and 98-014-01. These LERs reported problems with 4 penetration seals. These problems were resolved by the licensee.

6.2 Wolf Creek Nuclear Generating Station

6.2.1 Operating Experience

In December 1984, the licensee for Wolf Creek Nuclear Generating Station (WCNGS) issued a nonconformance report because 22 penetration seals lacked document traceability. The licensee completed corrective actions in 1985. Later, in early 1987, B&B Promatec Corporation (Promatec), Houston, Texas, the penetration seal installation contractor, notified the NRC that of 40 seals inspected, the silicone foam material in 20 showed voids and shrinkage. The problems had involved installation methodology, inadequate quality control (QC) methods, and rapid, chemically induced, expansion of the silicone foam material. The licensee issued LER 87-010 on February 6, 1987. This problem affected several other nuclear plants. Promatec informed the industry of the problems and submitted a Part 21 notification. The NRC issued IN 88-56 to advise licensees of the problems discovered at Wolf Creek.

In 1987, the licensee established a task force to develop a corrective action plan. The inspection plan covered the removal of damming boards and inspection of accessible foam penetrations. The scope of the program included inspections of more than 1700 silicone foam penetration seals. As a result of the inspections, the licensee repaired more than 600 seals during 1987. Since 1987, the licensee has found only minor problems during routine inspections, and the licensee addressed these promptly.

6.2.2 OI Investigation

In September 1988, the NRC Office of Investigations (OI) in Region IV initiated an investigation to determine if company officials at Promatec or WCNGS knowingly and intentionally failed to notify the NRC in 1984 and 1985 about the defective seals. In May 1987, Promatec had submitted a 10 CFR Part 21 report to the NRC, which stated that some silicone foam fire barrier penetration seals installed by Promatec at WCNGS did not meet minimum specifications. During replacement of damaged fire-resistant boards, WCNGS personnel found voids, shrinkage, and lack of fill in approximately 25 percent of the seals.

The OI investigation revealed that both Promatec and Kansas Gas & Electric (KG&E) became aware in 1983 of a similar problem with silicone seals at Callaway Nuclear Power Plant, also installed by Promatec. However, a different method of installation, a two-stage damming process, was utilized at WCNGS. Following the discovery of the problem at Callaway, Promatec conducted two seal reinspections at WCNGS. KG&E rejected the results of the first of these as too limited and indicative of a potential problem similar to the problem encountered at Callaway. The scope of the reinspection was expanded; the second reinspection led Promatec to conclude that there was a less than 2-percent rejection rate of these seals from shrinkage and voids. KG&E accepted the results of this reinspection and concluded that the problem at WCNGS was minor and not indicative of the problem found at Callaway.

On the basis of its investigation, OI concluded that the problem with the seals at WCNGS was generic, inherent both in the material and in the cable tie inspection method utilized at the time the seals were installed. OI concluded that the silicone material shrinks and expands depending on temperature changes and that it is difficult to install seals so as to ensure a complete fill, even utilizing the stage damming method of installation. OI also concluded that the inspection method used at WCNGS was inaccurate and could not reveal all voids, gaps, or missing fill in the seals.

From the time the seals were first installed, KG&E was aware of the seal inspection method used by Promatec. This was the acceptable method of inspection used by all sealing contractors at the time. Although KG&E knew about Callaway's problems, and was questioned by an American Nuclear Insurers (ANI) inspector and by the NRC regarding the adequacy of the inspection method, it took no steps to change to a visual inspection of the seals.

OI concluded that its investigation did not find evidence that KG&E or Promatec personnel were aware of specific problems at WCNGS and willfully failed to notify the NRC, as required by 10 CFR 50.55(e). OI also concluded that there is a potential for similar problems at any nuclear plant that utilized silicone foam seals and the method of inspection used at WCNGS, regardless of who installed the seals.

6.3 Salem Nuclear Generating Station

Fire barrier penetration seals have been inspected at least three times at the Salem Nuclear Generating Station. NRC Inspection 93-80 was an Appendix R inspection in which the licensee's penetration seal inspection program was evaluated. The inspection procedure was reviewed and the latest surveillance report was reviewed. The licensee inspects 10 percent of the fire barrier penetration seals every 18 months. If one failure is found, then an additional 10-percent sample is inspected until no more failures are identified. No failures were noted in the surveillance that was reviewed. The inspectors also reviewed the licensee's response to IN 88-56. The licensee's silicone foam seals were installed without the use of damming boards, making it very easy to detect voids or gaps.

Penetration seals were inspected again as a restart issue for Salem Inspection Report 96-10. The licensee had completed a 100-percent inspection and evaluation of all fire-rated penetration seals in 1992. The inspectors reviewed the design analyses of various types of penetrations and verified that the licensee's penetration seal details were representative of the tested seals, and that seals were bounded by acceptable fire endurance tests. The inspector concluded that the quality and configuration of penetration seals were acceptable.

The NRC recently inspected Salem's corrective actions to satisfy 10 CFR Part 50, Appendix R, Sections III.G, and III.L (Inspection Report 97-09). The inspectors compared "as built" penetration seals to the fire endurance test configurations to verify that as-built configurations were qualified by appropriate

fire endurance tests. The inspectors opened an inspection followup item (IFI) for as-built drawings, which did not identify important parameters with respect to cable fill and its thermal mass, and the maximum free area of unsupported penetration seal installed within the penetration.

Overall, the inspectors concluded that test specimens of the seals adequately represented and supported qualification of the as-built seal designs that were reviewed. The inspectors also concluded that the licensee's engineering analysis methods were adequate.

The licensee's staff has not identified any significant problems at Salem regarding penetration seals.

6.4 Millstone Nuclear Power Station

In LERs 93-006, 93-006-01, and 94-035, Millstone Nuclear Power Station reported penetration seal discrepancies. These LERs addressed unsealed penetrations found by the licensee. The staff reviewed LER 93-006 in NRC Inspection Report 93-19. The inspector reviewed the licensee's actions in response to the discovery of the missing seals, and reviewed the surveillance procedure that the licensee uses to inspect seals. The inspector noted that the procedure was adequate to enable proper inspection of the seals. The inspector noted that Unit 1 had identified only six other missing seals since 1990 through the seal surveillance program. This indicates that unsealed penetrations do not seem to be a programmatic concern at Millstone.

6.5 Maine Yankee Atomic Power Plant

6.5.1 NRC Inspection

From June 26–30, 1995, NRC Region I staff conducted a fire protection inspection at Maine Yankee Atomic Power Plant. The inspection is documented in NRC Inspection Report 50-309/95-15, which was transmitted to Maine Yankee Atomic Power Company (the licensee for the Maine Yankee plant) by letter dated September 20, 1995.

The inspector reviewed the fire barrier program to verify the adequacy of penetration seal installation, qualification, and inspection activities. This review also assessed the appropriateness of acceptance criteria established for penetration seals to validate operability and degradation that could prevent fire barriers from providing effective separation during a fire. The inspector concluded that the licensee's procedures for seal inspections, and the training program for seal inspectors, were good for maintaining proper seal configuration and for early detection of degraded conditions. These actions were found to provide a defense against the propagation of fire to adjacent plant areas.

The inspector reported that Maine Yankee relied on Insulation Consultants & Management Services, Incorporated (ICMS), to install the original penetration seals. The licensee informed the inspector that it had reviewed its purchase order information and project files and found that it did not apply any in-house quality control review for the ICMS fire barrier installation work. The licensee could not find the qualification and test reports completed by ICMS to support the seal installations, including fire and pressure test reports and qualification of seal installers. Therefore, the inspector could not verify the qualification of the penetration seals installed at Maine Yankee.

The inspector opened an unresolved item regarding the acceptability of penetration seal qualification, testing, and installer qualifications.

6.5.2 Licensee Event Reports

After the NRC staff fire protection inspection, the licensee conducted a scoping study in preparation for fire barrier penetration seal walkdowns. By letter dated July 29, 1996, the licensee submitted LER 96-017, "Fire Barrier Penetration Seal Discrepancy." The licensee reported that, during the scoping study, it found fire barrier wall penetration seals that did not have damming material in the proper location. On the basis of these findings, the licensee examined its criteria for penetration seals and conducted a technical review of its penetration seal design parameters. The licensee found discrepancies between available test reports and procedural guidance, and the in-plant penetration seal configurations. In response to the discrepancies, the licensee implemented compensatory fire watches and developed a corrective action program. The planned corrective actions were (1) determining why the discrepancies were not found during previous reviews; (2) evaluating the adequacy of procedures, test reports, acceptance criteria, and field inspections;

(3) evaluating the adequacy of existing seal configurations; and (4) inspecting all fire barrier penetration seals.

By letter dated August 28, 1996, the licensee submitted Revision 1 to LER 96-017. The licensee reported that it had found three additional types of deficiencies: (1) inadequate thickness of silicone foam, (2) temporary seals that were not upgraded to permanent seals for an indeterminate period, and (3) one seal for which the expected pipe movement exceeded the design rating of the seal.

6.5.3 Staff Followup

During a telephone conference on May 14, 1997, Office of Nuclear Reactor Regulation (NRR) and Region I staff obtained detailed information from the licensee regarding the seal problems found and the corrective actions. In addition, during the week of May 12, 1997, NRR staff reviewed and observed the problems found at Maine Yankee and the licensee's corrective actions.

The penetration seals at Maine Yankee were installed around 1978. Most of the original seals used silicone foam. Since the original installation, the licensee has visually inspected all the seals at each refueling outage.

During the inspections and walkdowns that were documented in LER 96-017-01, the licensee found that more than a thousand seals required further evaluation (including destructive examination); about a thousand other seals had defects; and a small number of seals had no defects. The licensee found seals with inadequate thickness (the predominant problem), foreign materials in seals, no damming material, and the wrong seal material installed. Although the licensee's design criteria specified a minimum seal thickness of 7 inches, the average seal thickness was 5 to 6 inches, and some seals were only 2 to 3 inches thick. Although the licensee once planned to repair and replace the seals with silicone foam and silicone elastomer, the licensee has since certified permanent cessation of power operation and is now proceeding to decommission the facility.

The licensee informed the staff that it believes that the installation deficiencies occurred because the quality assurance and quality control procedures used by the installation contractor during original seal installation were inadequate. The licensee also informed the staff that it believes it took so long to discover the deficiencies because its inspection and surveillance procedures did not cover all important penetration seal attributes (e.g., the presence of damming material was not a critical attribute) and because training was insufficient. The licensee has completed a major rewrite of its procedures.

The staff issued Information Notice (IN) 97-70, "Potential Problems With Fire Barrier Penetration Seals," on September 19, 1997, to tell industry of the problems found at Maine Yankee. As mentioned above, the licensee has since decided to shut the plant down permanently.

6.5.4 Conclusions on Maine Yankee Operating Experience

In NUREG-1552, "Fire Barrier Penetration Seals in Nuclear Power Plants" (July 1996), the staff stated that even though the overall condition of penetration seal programs in industry is satisfactory, it expects that plant-specific deficiencies may be found during future licensee surveillances and NRC inspections. Furthermore, the staff noted that licensees understand potential fire barrier penetration seal problems; industry consensus fire test standards are available and licensees adhere to them; and fire test results and qualified fire-resistant seal materials and designs are available. On these bases, the staff concluded that licensees have the means to correct problems, and staff oversight will continue to ensure corrections on a case-by-case basis. The penetration seal problems found by the NRC inspector at Maine Yankee and later reported by the licensee are consistent with the known types of problems, as previously documented by the staff in NUREG-1552. The reported problems do not indicate new trends.

6.6 Conclusions

LERs, NRC inspections, and plant-specific corrective action programs summarized above show that licensees knew and understood the fire-resistive capabilities of the penetration seal materials and configurations; potential penetration seal testing, design, installation, inspection, and maintenance problems; and possible remedies and corrective actions. These findings also indicate that the actions taken by the staff in 1988 and 1994 had increased industry awareness of possible penetration seal problems, leading industry to more comprehensive surveillance activities, maintenance practices, and corrective actions. To provide added assurance that

penetration seal deficiencies will be found, the staff revised the NRC fire protection core inspection module to provide specific inspection guidance to NRC inspectors.

7 REVIEW OF PLANT-SPECIFIC LICENSING BASES RELATED TO SECTION III.M OF APPENDIX R TO 10 CFR PART 50

7.1 Introduction

The following supplements information presented in Section 4 of NUREG-1552. On November 19, 1980, the U.S. Nuclear Regulatory Commission (NRC) published Appendix R, "Fire Protection Program for Nuclear Power Facilities Operating Prior to January 1, 1979," to Title 10 of the *Code of Federal Regulations* (10 CFR) Part 50, and a revised Section 50.48, "Fire protection," in the *Federal Register*. The revised Section 50.48 and Appendix R became effective on February 17, 1981. It is important to note that Appendix R is not a set of generically applicable fire protection requirements and that it applies only to plants that were operating before January 1, 1979.

Section III of Appendix R contains 15 subsections, lettered A through O, which specify requirements for nuclear power plant fire protection features. These requirements are divided into two categories. The first consists of those requirements that were backfit to facilities operating before January 1, 1979, regardless of whether or not the staff had previously approved alternatives to the requirements of those sections. These requirements are found in Section III.G, "Fire protection of safe shutdown capability"; Section III.J, "Emergency lighting"; and Section III.O, "Oil collection systems for reactor coolant pumps." The second category consists of requirements that were backfit on a plant-specific basis to the extent needed to resolve the "open" items of previous NRC staff fire protection reviews. An open item was defined as a fire protection feature that had not been previously approved by the NRC staff as satisfying the guidelines of Appendix A to Branch Technical Position (BTP) APCSB 9.5-1, as documented in a staff safety evaluation report (SER). Section III.M, "Fire barrier cable penetration seal qualification," of Appendix R was one such provision.

Section III.M states that penetration seal designs shall utilize only noncombustible materials[4] and shall be qualified by tests that are comparable to tests used to rate fire barriers. Section III.M contains the following acceptance criteria:

(1) Cable fire barrier penetration seal has withstood the fire endurance test without passage of flame or ignition of cables on the unexposed side.

(2) Temperatures recorded on the unexposed side are analyzed and the maximum temperature is sufficiently below the ignition temperature of the cable insulation.

(3) The fire barrier penetration seal remains intact and does not allow a projection of water beyond the unexposed surface during the hose stream test.

After it published Appendix R in the *Federal Register*, the staff sent letters to the licensees it applied to summarizing the open fire protection items and told each licensee which Appendix R requirements it had to comply with to resolve the items. Before the staff published NUREG-1552, Brookhaven National Laboratory (BNL), the staff's technical assistance contractor, reviewed these letters and found that 13 units had open items regarding fire barrier penetrations when Appendix R was published. They were:

Calvert Cliffs 1/2	Maine Yankee
Point Beach ½	Duane Arnold
Peach Bottom 2/3	Robinson 2
FitzPatrick	Pilgrim 1
Surry 1/2	

On the basis of BNL's review, the staff reported in NUREG-1552 that Section III.M of Appendix R applied to 13 nuclear power plants. In support of the review documented here, the staff again reviewed the licensing basis for the Appendix R plants and added Monticello and Vermont Yankee to the list of plants that may be required to comply with Section III.M of Appendix R. The staff then conducted a detailed review of the fire protection licensing bases for these

15 units. If the plants used silicone-based fire barrier penetration seal materials, which are classified as "combustible" when tested in accordance with ASTM Standard E-136,[5] the staff reviewed how the regulatory requirement of Section III.M of Appendix R that penetration seals utilize only noncombustible materials was addressed by the licensees. The findings of these reviews are documented below.

7.2 Plant-Specific Licensing Bases

7.2.1 Calvert Cliffs Nuclear Power Plant, Units 1 and 2

By letter dated November 24, 1980, the staff informed Baltimore Gas & Electric Company, the licensee for the Calvert Cliffs Nuclear Power Plant, Units 1 and 2, that the issue of ventilation and duct fire dampers was an open item. The issue of fire barrier penetration seals was not an open item. Therefore, Section III.M of Appendix R does not apply to the fire barrier penetration seals installed at Calvert Cliffs.

7.2.2 Duane Arnold Energy Center

Silicone-based penetration seal materials are installed in the plant.

In a letter of April 1, 1980, Iowa Light and Power Company, the licensee for the Duane Arnold Energy Center, stated that the penetration fire stops were conservatively designed and provided an adequate margin of safety for the plant fire protection design. In a letter of November 24, 1980, the staff informed the licensee that the tests described in its letter of April 1, 1980, did "not substantiate the fire resistance of the penetration seals installed at the plant." The staff also stated that "[t]o meet the requirements of Section III.M of Appendix R to 10 CFR Part 50, the licensee should provide additional documentation to verify that the seals which were tested and passed were representative of those actually installed."

The licensee responded in a letter of February 4, 1981, in which it compared the fire barrier penetration seal configurations it tested to those installed in the plant, and claimed that the

---

[4] A technical assessment regarding the combustibility of silicone-based seal materials is presented in Section 5.8 of NUREG-1552.

[5] "Behavior of Materials in a Vertical Tube Furnace at 750 °C," a pass/fail combustibility test method accepted by the NRC.

information provided in previous correspondence was sufficient to close the open item regarding fire barrier penetration seals.

7.2.3 James A. FitzPatrick Nuclear Power Plant

Silicone-based penetration seal materials are installed in the plant.

In a letter of February 13, 1981, the staff transmitted to the Power Authority of the State of New York, the licensee for James A. FitzPatrick Nuclear Power Plant (FitzPatrick), a supplemental SER in which it concluded that the silicone elastomer penetration seals installed at FitzPatrick met the criteria of Section III.M of Appendix R and were, therefore, acceptable. The open item regarding fire barrier penetration seals at FitzPatrick was closed before the effective date of Appendix R. Therefore, Section III.M of Appendix R does not apply to FitzPatrick.

7.2.4 Maine Yankee Atomic Power Plant

Silicone-based penetration seal materials are installed in the plant.

In Section 6.5 of this report, the staff discusses Maine Yankee. The plant has been permanently shut down and is being decommissioned.

7.2.5 Monticello Nuclear Generating Plant

In a letter of November 24, 1980, the staff informed Northern States Power Company, the licensee for Monticello Nuclear Generating Plant, that the cable tray penetrations at the south wall of the pipe and cable tray penetration area do not have adequate fire stops or adequate penetration seals. An NRC review determined that the vertical cable trays that penetrate the fire barrier are not sealed to provide adequate 3-hour fire resistance. Therefore, in order to comply with Section III.M of Appendix R, the licensee needs to install penetration seals that have a 3-hour fire-resistance rating. On October 20–24, 1986, a team of Region III and NRR personnel performed an announced inspection to determine the licensee's implementation of and compliance with the applicable requirements of 10 CFR Part 50, Appendix R. In Inspection Report 50-263/86008 (DRS), the inspection team determined, "the licensee

does now meet Section III.M of Appendix R and this 'Open' item is now considered closed."

7.2.6 Peach Bottom Atomic Power Station, Units 1 and 2

Silicone-based penetration seal materials are installed in the plant.

In a letter of November 24, 1980, the staff informed Philadelphia Electric Company, the licensee for Peach Bottom Atomic Power Station, Units 1 and 2, that the issue of penetration seals represented an open item. By letter of November 14, 1986, the staff issued an exemption from the technical requirements of Section III.M of Appendix R to the extent that certain penetration seals contain combustible material. In the safety evaluation supporting the exemption, the staff stated that the penetration "seals which contain combustible materials will provide an equivalent level of protection to that required by Section III.M of Appendix R." In the exemption, the staff stated that "the application of the regulation in this particular circumstance is not necessary to achieve the underlying purpose of the rule. Additionally, compliance with Section III.M concerning the subject seals would result in costs that are significantly in excess of those contemplated when the regulation was adopted since it would result in the complete removal and total replacement of all seals in question."

7.2.7 Pilgrim Nuclear Power Station, Unit 1

Silicone-based penetration seal materials are installed in the plant.

In a letter of December 15, 1980, the staff transmitted to Boston Edison Company, the licensee for Pilgrim Nuclear Power Station, Unit 1, an SER closing an open item regarding fire barrier penetration seals. In that SER, the staff stated: "[t]he licensee's proposed upgrading of penetration seals will result in seals which meet the requirements of Section III(M) [sic] of Appendix R to 10 CFR 50 and, therefore, are acceptable." The open item regarding fire barrier penetration seals at Pilgrim was closed before the effective date of Appendix R. Therefore, Section III.M of Appendix R does not apply to Pilgrim.

7.2.8 Point Beach Nuclear Plant, Units 1 and 2

Silicone-based penetration seal materials are installed in the plant.

In a letter of November 24, 1980, the staff informed Wisconsin Electric Power Company, the licensee for Point Beach Nuclear Plant, Units 1 and 2, that the issue of penetration seals was an open item and that the licensee was required to comply with Section III.M of Appendix R. In a letter of January 22, 1981, the staff transmitted to the licensee a supplemental SER, in which it concluded that the penetration seals installed at Point Beach met the criteria of Appendix A to BTP APCSB 9.5-1 and were, therefore, acceptable. The open item regarding fire barrier penetration seals at Point Beach was closed before the effective date of Appendix R. Therefore, Section III.M of Appendix R does not apply to Point Beach.

7.2.9 H.B. Robinson Steam Electric Plant, Unit 2

Silicone-based penetration seal materials are installed in the plant.

In a letter of November 24, 1980, to Carolina Power and Light Company, the licensee for H.B. Robinson Steam Electric Plant, Unit 2, the staff stated that to meet Section III.M of Appendix R to 10 CFR Part 50, "the licensee should provide cable penetration seals which utilize only noncombustible materials and should be qualified by tests that are comparable to those used to rate fire barriers." In a letter of November 25, 1983, the staff issued an exemption from the technical requirements of Section III.M of Appendix R to 10 CFR Part 50, to the extent that the acceptance criteria for penetration seal qualification required that the temperatures recorded on the unexposed side of the seal be below the cable insulation ignition temperature. Neither the exemption nor its supporting safety evaluation addressed the fact that the penetration seals used combustible materials.

7.2.10 Surry Power Station, Units 1 and 2

In a letter of November 24, 1980, to Virginia Electric and Power Company, the licensee for Surry Power Station, Units 1 and 2, the staff stated that "[t]o meet the requirements of Section III.M of Appendix R to 10 CFR 50, the licensee should upgrade all unsealed or inadequately sealed penetration openings to provide a 3-hour ASTM E-119 fire rated penetration seal where the fire rating of the barrier penetrated would be 3 hours." In a letter of December 18, 1980, the staff transmitted to the licensee a supplemental SER in which it concluded that the penetration seals installed at Surry met the criteria of Appendix A to BTP APCSB 9.5-1 and were, therefore, acceptable. The open item regarding fire barrier penetration seals at Surry was closed before the effective date of Appendix R. Therefore, Section III.M of Appendix R does not apply to Surry.

7.2.11 Vermont Yankee Nuclear Power Station

In a letter of January 13, 1978, the NRC issued Licensing Amendment 43 to Vermont Yankee Nuclear Power Station's operating license. In this amendment, the NRC identified Item 3.1.8, "Cable penetrations do not have a fire rating and do not provide adequate protection." In a letter of November 24, 1980, to Vermont Yankee Nuclear Power Corporation (VYNPC), the licensee for Vermont Yankee, the staff again noted that Item 3.1.8 was unresolved owing to the lack of supporting qualification tests. In a letter of December 19, 1980, to the NRC, VYNPC stated: "Vermont Yankee intends to maintain its commitment to provide 3-hour rated fire barrier penetration seals."

In a letter of December 31, 1980, Region I followup inspection 50-271/80-18 of Vermont Yankee fire barrier penetration seals, three inspection items were opened concerning the original "Item 3.1.8, Fire Barrier Penetrations." The open items were 80-18-01, an untested configuration; 80-18-02, questions on materials used to construct the penetration seals; and 80-18-03, a commitment to replace/upgrade existing penetration seals.

In a letter of December 23, 1981, a Region I Inspector reviewed open item 80-18-02, found the licensee actions acceptable, and closed the item. In a letter of April 22, 1982, Region I Inspectors reviewed open items 80-18-01 and 80-18-03, found the licensee actions acceptable, and closed the open items. Additionally, in an internal NRC memorandum dated April 16, 1982, to Thomas Novak, Assistant Director for Operating Reactors, from William Johnson, Assistant Director of Materials and Qualifications Engineering, Johnson stated: "open item 3.1.8 is now considered closed based on VYNPC's commitment to comply with Section III.M of Appendix R."

7.3 Summary

On the basis of its review of letters that the staff sent to the licensees of plants that were operating before January 1, 1979, after Appendix R was approved but before it became effective, it appeared that Section III.M of Appendix R applied to 15 nuclear power plants. However, on the basis of the detailed review summarized above, the staff found that Section III.M of Appendix R applied to Duane Arnold, H.B. Robinson 2, Maine Yankee, Monticello, Peach Bottom 2/3, and Vermont Yankee. Of these plants, the staff has granted exemptions for H.B. Robinson 2 and Peach Bottom 2/3. On the basis of its review of docketed information, the staff could not determine how the penetration seal open items were resolved at Duane Arnold and Maine Yankee. Because the licensee has permanently shut down Maine Yankee and is currently decommissioning it, the staff will not pursue this issue at Maine Yankee. The other plants discussed above, FitzPatrick, Pilgrim, Point Beach 1/2, and Surry 1/2, resolved the penetration seal open item before the effective date of Appendix R. Therefore, Section III.M of Appendix R does not apply to these plants.

8 RECOMMENDATIONS IN THE FINAL STAFF REPORT

8.1 Introduction

In SECY-96-146, "Technical Assessment of Fire Barrier Penetration Seals in Nuclear Power Plants" (July 1, 1996), the staff informed the Commission that the Office of Nuclear Reactor Regulation (NRR) had completed the subject assessment and forwarded to the Commission a copy of its final report entitled, "Technical Assessment of Fire Barrier Penetration Seals in Nuclear Power Plants" (June 14, 1996). In its final report, the staff recommended the following:

(1) Revise the NRC fire protection guidance documents to reflect the current National Fire Protection Association (NFPA) position on testing laboratories.

(2) Remove the noncombustibility criterion from Appendix R to 10 CFR Part 50 and Standard Review Plan (SRP) Section 9.5.1.

(3) Develop and issue guidance for comparing fire test configurations to as-built configurations.

(4) Make this technical assessment report available to the general public and industry by placing it in the NRC Public Document Room and issuing an information notice publicizing its availability.

In its final report, the staff also noted that it was preparing the new Fire Protection Functional Inspection (FPFI) Program that it had described in SECY-95-034, "Status of the Recommendations Resulting from the Reassessment of the NRC Fire Protection Program" (February 13, 1995). The staff stated that it would present guidance for inspecting fire barrier penetration seal programs in the FPFI procedures and guidelines for use by NRC inspectors on an as-needed basis.

8.2 Status

8.2.1 Recommendations 1, 2, and 3 (Pending)

Recommendations 1, 2, and 3 involved revising the NRC fire protection regulation (Appendix R) and review guidance (SRP). In its final report on penetration seals, the staff indicated that implementation of the recommendations would be useful to the industry, but did not identify technical or safety bases that justified an immediate need to implement them.

The NRC staff, under the Regulatory Improvements Program, is considering a performance-based, risk-informed fire protection regulation. After the staff issued its final report "Technical Assessment of Fire Barrier Penetration Seals in Nuclear Power Plants" (June 14, 1996), it issued several Commission papers regarding fire protection rulemaking. Most recently, in SECY-98-058, "Development of a Risk-Informed Performance-Based Regulation for Fire Protection at Nuclear Power Plants," March 26, 1998, the staff provided rulemaking options for a performance-based, risk-informed fire protection regulation; proposed to develop a comprehensive regulatory guide for reactor fire protection; and proposed to revise Section III.M of Appendix R to 10 CFR Part 50 to resolve the combustible penetration seal issue (see Section 5.8 of NUREG-1552).

In a staff requirements memorandum of June 30, 1998, the Commission directed the staff to develop the comprehensive regulatory guide and to pursue rulemaking to amend Section III.M of Appendix R to eliminate the requirement that penetration seal materials be noncombustible. Later, in a letter of July 20, 1998, from R.L. Seale, Chairman, ACRS, to Chairman Jackson, the ACRS stated its agreement with the Commission direction to amend Section III.M of Appendix R. The staff will implement the Commission's direction. This will satisfy the intent of Recommendations 1, 2, and 3.

8.2.2 Recommendation 4 (Complete)

In July 1996, the staff published NUREG-1552, "Fire Barrier Penetration Seals in Nuclear Power Plants." This action completed Recommendation 4.

8.2.3 FPFI Program (Complete)

The staff is currently using its FPFI procedures to conduct the pilot FPFI program. The NRC's routine fire protection inspection procedures are in the NRC Inspection Manual, Inspection Procedure 64704, "Fire Protection Program." In September 1997, the staff revised these procedures to provide more specific guidance for inspecting the seals and establishing their functionality.

9   PUBLIC COMMENTS ON DRAFT NUREG-1552, SUPPLEMENT 1

On July 13, 1998, the staff noticed in the *Federal Register* (Volume 63, Number 133) that it was accepting public comments on Draft NUREG-1552, Supplement 1. The staff also made the report available on the World Wide Web at the NRC website. During the public comment period, the staff received two letters in response to the draft report. In a letter dated September 11, 1998, the Nuclear Energy Institute (NEI) stated agreement with the conclusions of this report. In a letter dated September 16, 1998, the Nuclear Information and Resource Service (NIRS) stated disagreement with the conclusions of the report. Neither of the letters included new technical or safety information. Therefore, the comments did not result in changes to this report. These letters are part of the public record and are available at any NRC Public Document Room.

10   CONCLUSIONS

Since the fire at Browns Ferry Nuclear Plant in

March 1975, nuclear power plant licensees have made significant improvements in their fire protection programs. These improvements, especially the adoption of the defense-in-depth concept of echelons of fire protection, have reduced both the probability and the potentially adverse consequences of nuclear power plant fires. Using documented industry operating experience, the staff carefully and objectively evaluated issues associated with fire barrier penetration seals. The staff considered the potential safety and risk significance of potential penetration seal deficiencies and the use of compensatory measures for any potential degradation in the fire protection effectiveness of seals.

For the reasons discussed in Sections 3 through 5, the staff considers that the relative safety significance of the subject fire barrier penetration seal concerns is low. Even assuming that certain fire barrier penetration seals are deficient, it does not follow that the deficiencies indicate the absence of adequate protection. The Commission has explained that

> [W]hile it is true that compliance with all NRC regulations provides reasonable assurance of adequate protection of the public health and safety, the converse is not correct, that failure to comply with one regulation or another is an indication of the absence of adequate protection, at least in a situation where the Commission has reviewed the noncompliance and found that it does not pose an "undue risk" to the public health and safety.[6]

The failure to have fire barrier penetration seals that meet the criteria specified by the NRC fire protection guidance documents does not necessarily indicate that a plant is unsafe.

On the basis of everything it found and considered, it is the staff's judgment that, overall, the issue of potential fire barrier penetration seal deficiencies does not adversely affect safety. For the reasons given in this paper, typical penetration seal deficiencies do not necessarily equate to a lack of adequate protection or result in undue risk to public health and safety.

On the basis of the reassessment documented here,

---

[6]Ohio Citizens for Responsible Energy, DPRM 88-4, 28 NRC 411 (1988).

the staff concluded that the actions it took in 1988 and 1994 to alert licensees to potential penetration seal problems increased industry awareness of such problems and resulted in more thorough surveillances, maintenance, and corrective actions.

The staff also concluded that the general condition of penetration seal programs in industry is satisfactory.

The staff will continue its reviews and inspections of penetration seal programs. The staff expects that plant-specific deficiencies may occasionally be found during licensee surveillances and NRC reviews and inspections. However, potential penetration seal problems are understood; industry consensus fire test standards are available and are followed; and fire test results and qualified fire-resistant seal materials and designs are available. Therefore, licensees have the means to correct problems, and continued staff oversight will continue to ensure corrections on a case-by-case basis. In addition, the fire protection defense-in-depth concept provides reasonable assurance that deficiencies will not present an undue risk to public health and safety before they are found and corrected.

The results of this assessment, which used information that the staff had not considered in the evaluation documented in NUREG-1552, "Fire Barrier Penetration Seals in Nuclear Power Plants," have reinforced the staff's earlier conclusion that RTV silicone foam penetration seals like other types of penetration seals installed in US nuclear plants, provide reasonable assurance that a fire in a specific fire area or zone will be confined to the area of origin.

During the 454th meeting of the Advisory Committee on Reactor Safeguards (ACRS), July 8-10, 1998, the staff presented the results of the assessment documented in this supplement to NUREG-1552 to the ACRS. The ACRS provided its views regarding the efforts of the NRC staff and the nuclear industry to resolve issues related to fire barrier penetration seals in a letter of July 20, 1998, from R.L. Seale, Chairman, ACRS, to Chairman Jackson. The ACRS noted that it is clear that, overall, the NRC staff and the licensees have the issues of fire barrier penetration seals well in hand and that the efforts of the staff and the licensees have been successful in addressing the problems of the past.

In sum, it is the staff's opinion that continued licensee attention to existing penetration seal programs and continued NRC inspections are adequate (1) to ensure that penetration seal problems are discovered and resolved and (2) to maintain public health and safety. To provide added assurance of this, during the assessment documented in this report, the staff issued Information Notice 97-70, "Potential Problems With Fire Barrier Penetration Seals," September 19, 1997, and revised the NRC fire protection core inspection module to provide more specific inspection guidance to NRC inspectors regarding fire barriers and fire barrier penetration seals. The staff will continue to assess new information regarding penetration seals for new insights and appropriate opportunities for additional actions by the staff or the industry.

Appendix D

Acronyms and Abbreviations

BNL   Brookhaven National Laboratory

CDF   core-damage frequency
CDR   construction deficiency report

DRS   division of reactor safety

ICMS   Insulation Consultants & Management Services, Incorporated
IFI   inspection followup item

KG&E   Kansas Gas & Electric

OI   Office of Investigations (NRC)
ORNL   Oak Ridge National Laboratory

PVC   polyvinyl chloride

RTV   room temperature vulcanizing

SER   safety evaluation report
SRM   staff requirements memorandum

URI   unresolved issue

VYNPC   Vermont Yankee Nuclear Power Corporation

WCNGS   Wolf Creek Nuclear Generating Station

Appendix F

Licensee Event Reports Submitted by Year
(1987 Through September 1998)

Year	Number of Sites	Number of LERs	Number of Supplements
1987	12	16	3
1988	9	12	4
1989	12	14	9
1990	8	11	5
1991	7	8	10
1992	3	8	8
1993	7	8	6
1994	6	6	5
1995	4	4	3
1996	5	5	1
1997	4	3	3
1998	4	5	3
TOTAL	46	100	62

NUREG-1552, Supp. 1

Appendix G

Summary of Reported Problems
(1987 Through September 1998)

Reported Problems	Number of Occurrences												Subtotal	Total
	87	88	89	90	91	92	93	94	95	96	97	98		
Penetrations unsealed	10	6	6	4	1	3	4	1	1	-	1	1	38	
Seal breached and not repaired	4	1	1	2	-	2	-	-	1	-	-	-	11	
Internal conduit seal not installed	1	1	3	1	-	1	-	2	-	-	-	-	9	
Total - Seal Not Installed or Breached	15	8	10	7	1	6	4	3	2	-	1	1		58
Voids, gaps, splits, shrinkage, cell structure	1	4	3	4	2	2	-	2	-	-	-	1	19	
Inadequate seal thickness	1	2	1	-	-	3	2	2	-	1	-	2	14	
Seal not properly installed	-	-	1	1	2	2	2	2	-	1	-	2	13	
Incorrect seal material installed	1	2	1	-	1	-	-	1	-	-	1	-	7	
Temporary seal not replaced	1	-	1	-	-	1	1	-	-	1	-	-	5	
Inadequate seal repair	-	1	-	-	1	1	-	-	1	-	-	1	5	
Total - Seal Not Properly Installed	4	9	7	5	6	9	5	7	1	3	1	6		63
Total - Inadequate Documentation	1	5	2	3	1	-	1	1	1	3	1	-	19	19
Seal degraded or damaged	2	2	1	-	1	3	-	-	1	1	-	-	11	
Missing or damaged damming boards	1	1	1	-	-	1	-	1	-	1	-	-	6	
Total - Seal Degraded or Damaged	3	3	2	-	1	4	-	1	1	2	-	-		17
Totals	23	25	21	15	9	19	10	12	5	8	3	7		157

Appendix H

Summary of Licensee Event Reports
(1987 Through September 1998)

(Appendix R plants (plants operating prior to January 1, 1979) are shown in **bold** font.)

		1987	
PLANT	**LER NO.**	**ACCESSION NO.**	**REPORT**
ANO 2	87-001-00	8703180073	2 conduits missing internal seals.
FitzPatrick	87-011-00	8709020094	224 out of a total of 16,000 penetrations found unsealed.
	87-011-01	8802030335	Updated 87-011-00. Installation specification, surveillance procedures revised.
Fort St. Vrain 1	87-006-00	8704160030	Unsealed penetrations and degraded seals.
	87-006-01	8705180247	Updated 87-006-00.
Monticello	87-011-00	8705260063	1 unsealed penetration.
Nine Mile Point 2	87-016-00	8703310063	1 penetration sealed with incorrect seal material. Similar seals inspected.
	87-016-01	8707010536	Unsealed penetrations and breached seal.
	87-018-00	8704150327	1 breached seal.
Quad Cities 1/2	87-028-00	8803080281	Several damaged seals, several unsealed penetrations, and 7 inadequate temporary seals.
River Bend Station	87-021-00	8711170189	2 unsealed penetrations.
Salem 1/2	87-007-00	8706150188	1 unsealed penetration.
Susquehanna 1	87-011-00	8705050296	1 unsealed penetration.
TMI-1	87-003-00	8705080327	1 unsealed penetration.
WNP2	87-004-00	8705130234	Design drawings were incomplete, 2 unsealed penetrations, and 1 seal not included in surveillance procedure.
	87-029-00	8710220153	1 seal not repaired after breaching to remove cables.
	87-030-00		Penetrations not sealed.

1987 (continued)			

PLANT	LER NO.	ACCESSION NO.	REPORT
Wolf Creek	87-001-00	8702100286	1 seal found breached.
	87-010-00	8703250035	Several seals found breached. Surveillance procedure enhanced, personnel trained.
	87-010-01	8707150537	Fire protection program to be upgraded. Nonconforming silicone foam seals found (missing or damaged damming boards, inadequate seal thickness, voids, shrinkage).
	87-010-02	8804050361	Final update of 87-010-00. Performed sample inspection program by removing damming boards from 40 seals; 13 rejected for insufficient foam thickness, 9 rejected for voids and shrinkage. Performed **100% inspection** (1700 seals). Repaired and reworked more than 600 seals.

1988			

PLANT	LER NO.	ACCESSION NO.	REPORT
Ginna 1	88-009-00	8811090368	Several degraded seals and seals with incorrect seal material found.
H.B. Robinson 2	88-018-00	8810070343	101 cable tray penetration seals inspected. 38 not sealed inside tray covers due to inadequate installation procedure. Procedures revised.
	88-018-01	8906190260	Updated 88-018-00.
McGuire 1	88-030-00	8811150235	Review conducted in response to **IN 88-04**. 96 seals declared inoperable due to lack of test documentation.
	88-030-01	89022700381	Updated 89-030-00. Seals qualified by test. Procedures improved.
Nine Mile Point	88-009-00	8804280564	Replaced by 88-009-01.
	88-009-01	9006180174	Task force formed and **100% seal inspection** initiated. 13 seals did not have adequate supporting documentation. Fire protection program enhanced.
	88-009-02	9008230138	14 seals did not have adequate documentation.

| | 1988 (continued) | | |

PLANT	LER NO.	ACCESSION NO.	REPORT
North Anna 1/2	88-007-00	8802290350	Eight fire barrier penetration seal breaches were identified. These breaches were repaired.
Oconee 1/2/3	88-005-00	8806270349	Review conducted in response to **IN 88-04**. **100% seal inspection** revealed 188 inoperable seals due to inadequate documentation. Procedures revised.
River Bend Station	88-009-00	8804050384	3 unsealed penetrations and one inadequate seal found.
	88-009-01	8805100011	1 unqualified penetration seal found.
	88-009-02	8808310152	Unsealed conduits, unsealed penetrations, breached seals, and incompletely sealed penetrations found.
Salem 1/2	88-013-00	8809140180	Several silicone foam seals did not conform to correct color and cell structure. Existed since original installation. Installation procedure revised. **100% of foam seals inspected** to verify compliance with installation criteria.
	88-014-00	8810040008	Purpose of LER was to report missed surveillance for inoperable penetration seals. Also, summarized seals inoperable because of degradation, wrong seal material, shrinkage, and unsealed penetrations.
Waterford 3	88-011-00	8806300078	1 seal found that did not conform to standard design.
	88-025-00	8811170093	Unsealed penetrations found.
	88-030-00	8812150039	**100% seal inspection.** Found unsealed penetrations, missing damming boards, and silicone foam seals with voids.
	88-030-01	8906050115	Updated 88-030-00. Damming boards removed from seals for inspection. Found 99 seals with voids, 123 seals that differed from typical design details, 17 seals that deviated from vendor requirements, and 19 unsealed penetrations.
	88-030-02	8907190362	Updated 88-030-00. Installation procedures changed.

Summary of Licensee Event Reports

1988 (continued)			
PLANT	LER NO.	ACCESSION NO.	REPORT
Waterford 3 (continued)	88-030-03	9109060034	Updated 88-030-00. 228 seals to be reworked.
WNP2	88-008-00	8805030155	11 inoperable seals due to unapproved configuration, inadequate seal thickness, seals improperly repaired. Updated seal database. **100% documentation review and seal inspection.**
	88-008-01	9302220125	Updated 88-008-00.
1989			
ANO 1	89-003-00	8903280098	2 penetrations sealed with unqualified material.
Big Rock Point	89-006-00	8908240314	Licensee initiated penetration seal verification program in response to **IN 88-04 and IN 88-56**. 1 seal breached and not repaired, 1 seal inadequately installed.
	89-006-01	9004130265	3 inadequate seals and 1 seal with a gap were found.
Calvert Cliffs 2	89-002-00	8904050315	Conduit missing internal seal.
	89-002-01	8911210052	Updated 89-002-00.
Clinton 1	89-006-00	8902230041	3 conduits missing internal seals.
Dresden 2	89-030-00	8911280062	1 unsealed penetration. Procedures improved.
Fort St. Vrain	89-014-00	8909250113	4 seals did not meet cell structure criteria.
	89-014-01	8912270289	Updated 89-014-00, 2 seals deleted from the LER.
Haddam Neck	89-001-00	8902070157	1 temporary seal found inoperable. Seal upgrade program conducted in response to **IN 88-04**.
	89-001-01	9101140199	Several unsealed penetrations found during seal upgrade program.

		1989 (continued)	

PLANT	LER NO.	ACCESSION NO.	REPORT
Monticello	89-001-00	8902080493	6 unsealed penetration found. **100% inspection** initiated.
	89-013-00	8908070189	Several unsealed penetrations found.
	89-013-01	9001100234	Updated 89-013-00. Inspection completed. No additional deficiencies found.
North Anna 1/2	89-003-00	8902140025	A void was discovered in one fire barrier penetration seal. A fire watch was put into place, and the void was then repaired.
Palisades	89-024-00	8912260122	Inspection conducted in response to **IN 88-04**. 1 unsealed penetration found.
River Bend Station	89-005-00	8903240060	Void found in 1 low-density silicone elastomer seal. Sample of similar seals inspected.
	89-010-00	8904260064	1 unsealed penetration and 4 conduits without internal seals.
	89-010-01	8906190263	Updated 89-010-00. Task force formed.
	89-010-02	8909080115	Updated 89-010-00.
	89-010-03	9008060246	Updated 89-010-00. Based on results of sample inspections, conducted **100% seal inspection**.
	89-010-04	9401060365	Completed program end of 1993. 3385 penetration seals inspected; 1961 found unacceptable. Reworked or reevaluated deficient seals. Deficiencies included: gouged or damaged damming material, shrinkage of silicone foam, inadequate seal thickness, cuts in boot material, and inadequate documentation.
	89-010-05	9409140061	Updated 89-010-00.
Seabrook	89-011-00	8910170274	3 unsealed pipe penetrations.
	89-011-01	8912270219	Updated 89-011-00. Initiated **100% seal inspection**, developed comprehensive seal program, clarified surveillance requirements.

		1989 (continued)	
PLANT	**LER NO.**	**ACCESSION NO.**	**REPORT**
Susquehanna	89-019-00	8907060047	Damaged seals determined to be inoperable. Consistent inspection and acceptance criteria developed.

		1990	
ANO 1	90-004-00	9007090045	1 unqualified penetration seal.
	90-004-01	9105160074	Small voids around grout joint.
	90-004-02	9204300230	In response to **IN 88-04**, found 2 seals not properly installed.
	90-017-00	9008200077	Void in large grout blockout seal.
	90-023-00	9012120354	1 unsealed penetration.
Fort Calhoun 1	90-022-00	9010170151	In response to **IN 88-04**, assessed and walked down **100% of seals**. Found about 460 of 3500 seals may be inoperable because documentation did not exist or installed configurations did not match documentation.
	90-022-01	9101090184	Updated 90-022-00. Found 92 more potentially inoperable seals.
	90-022-02	9102120021	Updated 90-022-00. Found more potentially inoperable seals and resolved others. Final count of potential inoperable seals due to lack of documentation was 441 out of 3500. The licensee performed evaluations, repaired, and replaced seals. Upgraded procedural controls and drawings.
H.B. Robinson 2	90-003-00	9002220099	Missing internal conduit seal.
	90-008-00	9006050277	1/4" plastic tube found passing through (breaching) a seal.
	90-010-00	9002220099	1 unsealed penetration.
	90-010-01	9103270201	Performed **100% inspection**, 14 additional inoperable seals found.
Monticello	90-009-00	9008280179	Seal breached and not resealed.

1990 (continued)

PLANT	LER NO.	ACCESSION NO.	REPORT
Palo Verde 1/2/3	90-009-00	9010310125	Performed **100% inspection** of Unit 2, found about 256 questionable seal attributes out of 2000 examined. Deficiencies included unsealed penetrations, seal shrinkage, improperly installed seals, and gaps in damming materials.
	90-009-01	9208200192	Performed **100% inspection** of Units 1 and 3. Found about 1437 questionable seal attributes out of more than 10,000 examined. Deficiencies included unsealed penetrations, seal shrinkage, improperly installed seals, and gaps in damming materials.
Trojan	90-022-00	9007230142	In response to **IN 88-56,** found silicone foam seals with splits.
	90-022-01	9012060223	Destructive testing revealed 17 similar seals with splits.
Waterford 3	90-019-00	9101150362	Removed penetration seal around HVAC damper as a part of modification and did not replace.
	90-019-01	9103040377	Updated 90-019-00. Found 1 additional unsealed penetration.
	90-019-02	9109190291	Updated 90-019-00.

1991

PLANT	LER NO.	ACCESSION NO.	REPORT
ANO 2	91-016-00	9110250001	Seal not installed properly (filled with rags rather than grout).
Big Rock Point	91-001-00	9102200140	Voids found in 3 seals in response to **IN 88-56**.
	91-001-01	9103260311	8 more seals found with voids.
FitzPatrick	91-024-00	9112170535	7 penetrations sealed with incorrect material.
	91-024-01	9403230046	Performed **100% inspection**. Deviations from design were found in 39% of 7200 seals inspected. 15% required cosmetic repairs. Problems included: inadequate seal thickness, installation, or seal material, unsealed penetrations, voids, holes, edge curl, and separation of foam. All seals were restored to design condition or evaluated.

1991 (continued)			

PLANT	LER NO.	ACCESSION NO.	REPORT
Monticello	91-021-00	9111050217	Seal damaged due to pipe movement.
Peach Bottom 2	91-013-00	9106190190	2 seals contained voids and uncured sealant material.
Point Beach 1	91-007-00	9107300239	2 seals left inoperable after design modification.
Sequoyah 1	91-013-00	9107030303	Improperly installed seal around a conduit
	91-013-01	9108050172	Updated 91-013-00.
	91-016-00	9108190108	9 mechanical seals inoperable due to pipe movement.
	91-016-01	9202140203	Schedule update.

1992			

PLANT	LER NO.	ACCESSION NO.	REPORT
Duane Arnold	92-003-00	9203190032	1600 seals inspected. 1 penetration found unsealed since design modification. Program improvements made to minimize likelihood of recurrence.
	92-007-00	9206150398	6 penetrations unsealed since original plant construction. Found during first time inspection using new, enhanced inspection program.
	92-007-01	9208040177	Updated 92-007-00. Improved inspection schedule.
Haddam Neck	92-008-00	9203270186	1 seal inoperable. Silicone foam had been removed and replaced with ceramic fiber.
Trojan	92-006-00	9203090105	2 seals missing damming boards and inadequate silicone foam thickness since original installation (1979). Corrective actions included inspecting all similar seals.
	92-006-01	9205110198	Inspection of similar seals found 1 additional seal without damming board.
	92-011-00	9206080031	1 seal not repaired and 1 breached seal not resealed. Fire barrier inspection procedures were upgraded.
	92-026-00	9209300187	During 18-month surveillance found grout missing from 1 seal. Inspectors retrained.

1992 (continued)			

PLANT	LER NO.	ACCESSION NO.	REPORT
Trojan (continued)	92-026-01	9211030238	1" diameter hole found through a silicone foam seal.
	92-026-02	9211160031	1 seal with inadequate grout thickness and 1 grout seal damaged.
	92-026-03	9211300072	2 conduits did not have internal seals.
	92-026-04	9301050162	4 seals found with inadequate thickness of silicone foam and 1 seal with inadequate thickness of grout. Personnel retrained.
	92-026-05	9310250073	Updated 92-026-00. Degraded penetration seals resulted from personnel errors and inadequate procedural controls. Extensive procedural controls implemented.
	92-031-00	9211190123	1 grout seal degraded and inadequate grout thickness.
	92-034-00	9301250264	A small gap was found between a grout seal and the penetrating pipe. Two grout seals were degraded and 1 of these had inadequate grout thickness.

1993			

PLANT	LER NO.	ACCESSION NO.	REPORT
Brunswick	93-006-00	9304060055	During **100% inspection**, found 9 unqualified seals.
Haddam Neck	93-003-00	9305030266	Found 1 unsealed penetration and 1 seal with a temporary seal
Indian Point 3	93-029-00	9309240036	In response to **IN 88-04**, initiated seal inspection program. 2 seals found that did not conform to tested configuration.
LaSalle 1	93-009-00	9303290295	3 unsealed penetrations. Sample of penetrations inspected. No additional deficiencies found.
Millstone 1	93-006-00	9307200165	1 unsealed penetration found using improved inspection procedure.
Trojan	93-001-00	9302230261	1 unsealed penetration.
	93-002-00	9303180036	2 grout seals had inadequate thickness.

1993 (continued)			
PLANT	**LER NO.**	**ACCESSION NO.**	**REPORT**
Vermont Yankee	93-001-00	9301220246	In 1992, all seals containing insulated lines were declared indeterminate. Inspection revealed 1 penetration with inadequate seal thickness and 3 others that did not conform to design details. Licensee notified industry through Nuclear Network.
	93-001-01	9303090037	Updated 93-001-00. Boot seals to be used for some pipe penetrations.
	93-001-02	9307140180	Updated 93-001-00. All seals to be inspected using enhanced surveillance procedure. Design change implemented.
1994			
Cooper	94-008-00	9405240103	Improperly installed seal found. Seal was repaired.
Diablo Canyon 1/2	94-001-00	9403090054	Seals may not meet required fire rating due to lack of damming boards. All seals declared indeterminate. Program to qualify and repair seals.
	94-001-01	9408310118	Updated 94-001-00.
Maine Yankee	94-010-00	9408180131	2 conduits without internal seals found.
	94-010-01	9508290022	Conduit seals missing. The conduits were part of a new installation. They were sealed and inspected.
Millstone 2	94-035-00	9412060226	Breached/missing internal conduit seal. Seals installed.
Vermont Yankee	94-018-00	9501190145	2 seals degraded. One was missing caulk and the other had a 3/8" void in the brick and mortar seal.
	94-018-01	9506140431	Updated 94-018-00.
WNP2	94-002-00	9403230142	Due to an employee concern, licensee found original installation of seals, including written procedures, design configuration, and analysis less than adequate. Deficiencies included: inadequate thickness, PVC sleeves and seals that exceeded design specifications. Seals declared inoperable. Corrective actions included walkdowns, engineering evaluations, and establishing supporting fire test documentation.

1994 (continued)			
PLANT	**LER NO.**	**ACCESSION NO.**	**REPORT**
WNP2 (continued)	94-002-01	9407130092	Updated 94-002-01.
1995			
Calvert Cliffs 1	95-004-00	9509210118	3/4" gap (breach) found in a seal. Seal repaired, seal surveillance procedure upgraded.
Haddam Neck	95-001-00	9502230065	1 degraded grout seal and 1 unsealed penetration found. 18-month surveillance revealed 4 inoperable seals and 3 unsealed penetrations. **100% field walkdown** as corrective action.
	95-001-01	950808017	Updated 95-001-00.
Susquehanna	95-011-00	9511070336	Review of fire test reports revealed that hose stream test did not meet commitment. Staff inspected this issue January 1996.
Vermont Yankee	95-004-00	9505030454	Improperly repaired seal declared inoperable. Seal was repaired.
1996			
D.C. Cook 2	96-004-00	9604180325	Seal found degraded/damaged when a **100% seal inspection** was completed.
Diablo Canyon 1/2	96-011-00	9609170363	Epoxy grout seals untested and, therefore, outside design basis.
	96-011-01	9706040331	Reported qualification of epoxy grout seals by test.
Maine Yankee	96-017-00	9608060017	Fire barrier penetration inspection revealed seals missing damming boards, inadequate seal thickness, and temporary seals. No fire tests to support some configurations. Attributed to weaknesses in original installation QC, and surveillance procedures.
	96-017-01	9608060017	Updated 96-017-00.
Palisades	96-009-00	9608200212	Fire barrier evaluations not documented for two seals. Penetration seal program weaknesses noted. Commitment made to develop a design-basis document for fire barriers.

1996 (continued)			

PLANT	LER NO.	ACCESSION NO.	REPORT
Vermont Yankee	96-026-00	9611130511	Two seals improperly installed during original installation.
	96-026-01	9703280401	Updated 96-026-00.

1997			

Fermi	97-014-01	9804140118	Penetration seals were found missing from the auxiliary building wall and parallel turbine building wall. These walls were rated fire barriers. 16 penetrations were not sealed at the auxiliary building wall, and 4 were unsealed at the turbine building wall. The unsealed penetrations were sealed to comply with Appendix R.
St. Lucie	97-004-00	9707150008	Two-sided cable tray firestop was discovered to be missing ceramic fiber insulation between cables. All cable tray fire stops were declared inoperable, and will be upgraded accordingly.
	97-008-00	9709040179	15 penetration seals were declared inoperable as they could not be bounded by supporting fire tests. The seal manufacturer (Promatec) did not supply proper qualification documentation. Seals will be re-worked to meet applicable configuration drawings.

1998			

Clinton	98-021-00	9808250144	Cracks were discovered in a penetration seal during a "NUREG 1552" walkdown of penetration seals. Some cracks went completely through the seal. The licensee is inspecting other seals, repairing any that need repair and revising procedures. Licensee issued 10 CFR Part 21 notification concerning the seal material.

1998 (continued)

PLANT	LER NO.	ACCESSION NO.	REPORT
Prairie Island	98-003-00	9806300550	Penetrations were discovered that were not sealed. The openings were to be evaluated and then sealed according to plant procedures.
Vermont Yankee	98-001-00	9803020316	Following work potentially affecting a penetration seal, the seal was inspected and found not to conform to the tested configuration for a 3-hour seal. The grout seal was inadequately installed during the construction of the plant and was to be repaired. Other grout block-out seals were to be inspected.
	98-001-01	9805210006	Updated 98-001-00.
	98-008-00	9805060322	Penetration seal was found to have 3" of seal depth where 6" was required for a 3-hour rating (silicone elastomer). Root cause was determined to be inadequate QA/QC on original installation.
	98-008-01	9808240312	Updated 98-008-00.
	98-014-00	9806250116	VY discovered 2 non-conforming seals. One seal was inadequately repaired with 7" of silicone foam rather than the required 12" for a 3-hour rating. The other seal was improperly installed with 7" of foam rather than the required 12". The seals were repaired.
	98-014-01	9808180028	Updated 98-014-00.

Appendix I

NRC Inspections (March 1988 Through August 1998)

(Appendix R plants (plants operating prior to January 1, 1979) are shown in **bold** font.)

Plant	Report	Date	Scope	Findings	Summary
Beaver Valley 1 **Beaver Valley 2**	93-12, 93-13	07/02/93	Narrow	Minor	Licensee could not verify that eight internal conduit seals were installed. A fire watch was posted until the seals were installed per procedures.
Browns Ferry 1/2/3	89-28	09/15/89	Narrow	None	During a fire protection inspection, inspectors opened followup item 89-28-03 to track completion of penetration seals for electrical raceways and mechanical fire barrier penetrations. Inspectors found the licensee's penetration seal program to be acceptable.
	90-11	05/11/90	Narrow	None	Inspectors closed followup item 89-28-03 regarding installation of penetration seals.
	92-11	05/01/92	Narrow	None	Inspectors reviewed procedures for maintenance of fire barrier penetrations. Inspection results for fire rated barriers were also reviewed. No discrepancies were noted.
	95-60	12/12/95	Broad	None	Inspector reviewed typical mechanical, electrical conduit, and cable tray penetration seal installation procedures, drawings, details, quality control (QC) records, quality assurance (QA) records, engineering evaluations, and qualification test documentation. Inspector did not find any discrepancies.
	98-01	03/24/98	Broad	None	Inspectors reviewed the licensee's penetration seal program and determined that it was adequate. Licensee had evaluated numerous seals to demonstrate that they were adequate for their given applications. Licensee was performing 100% seal inspection every 18 months. This was considered a strength in the fire protection program.
Brunswick 1/2	92-31	10/26/92	Narrow	None	Inspectors observed performance of a portion of the periodic inspection of fire barrier seals. Inspectors noted the inspections were detailed, and that the licensee had initiated a re-inspection effort for fire barriers, which was seen as a positive initiative for self-identification and corrective action of fire barrier inspection program deficiencies. In addition, inspectors noted that penetration seals were acceptable during a general plant walkdown.
	93-08	03/25/93	Narrow	None	During a fire protection inspection, inspectors reviewed the licensee's fire barrier reinspection program and found it to be adequate.
	93-38	09/10/93	Narrow	None	Inspectors closed LER 92-12-01 which concerned inadequate fire barrier wall gap material. As part of the close-out actions, the licensee conducted a detailed review and inspection of fire barriers and penetration seals during a Unit 1 outage.
	97-07	06/20/97	Narrow	None	Inspectors noted that penetration seals were acceptable during a general plant walkdown.
	97-13	01/23/98	Broad	Minor	Inspectors reviewed a sample of silicone foam fire barrier penetration seals including seal design and testing. Inspectors opened IFI 97-13-04 to track missing penetration seal testing documentation.

Plant	Report	Date	Scope	Findings	Summary
Byron 1/2	92-007	04/13/92	Narrow	None	Inspector observed fire penetration seals while conducting a plant walkdown and did not observe any problems.
Callaway	94-012	12/06/94	Narrow	None	Inspectors noted that barrier seals in the plant were in generally good condition.
Calvert Cliffs 1/2	94-15	05/06/94	Narrow	None	Inspectors noted that the licensee has scheduled a review of all plant penetrations to verify the adequacy of the installations. Inspectors concluded that there were no safety-significant issues associated with the penetration seals.
	93-99	07/10/95	Broad	None	SALP report concluded that licensee lacked a formal engineering evaluation for qualification of certain fire barrier penetration seal materials.
	95-08	10/16/95	Broad	Minor	Cork expansion joints found to be inadequate fire barriers; polysulfide caulk found to be inadequate sealant for a fire-rated barrier. These deficiencies resulted in a Severity Level IV violation.
	96-201	05/06/96	Broad	None	The staff inspected the fire barrier penetration seal program and concluded that the licensee had an acceptable program. Ongoing licensee efforts to improve the penetration seal program were seen as positive.
Catawba 1/2	91-22	11/04/91	Narrow	None	During a fire protection inspection, inspectors did not identify any discrepancies in fire barrier penetration seal installations while on a plant walkdown.
	97-07	05/23/97	Broad	None	Inspectors reviewed licensee's evaluations and corrective actions related to IN 94-28, "Potential Problems with Fire Barrier Penetration Seals."
	98-07	07/27/98	Broad	Minor	Inspectors reviewed licensee's corrective actions for penetration seals that were found with gaps and lack of proper sealant material. Inspectors issued a non-cited violation for the noncompliance.
Comanche Peak 1/2	96-10	09/24/96	Narrow	None	Inspectors observed installation of a penetration seal and no discrepancies were noted.
	96-12	11/27/96	Broad	None	Inspector inspected silicone foam seals and verified that they were installed in the proper configuration and had adequate documentation to support a 3-hour fire rating.
Cooper	95-17	02/05/96	Narrow	None	Inspectors closed LER 94-008 regarding inoperable penetration seals.
Crystal River	92-18	10/01/92	Narrow	None	Inspectors reviewed fire barrier penetration technical specification requirements, including daily fire barrier breach reports.
	97-18	01/06/98	Narrow	Minor	Inspectors conducted an Appendix R inspection. Inspectors closed restart issues on penetration seals. Inspectors opened IFI 97-18-01 to track lack of documentation supporting the seal installations.

Plant	Report	Date	Scope	Findings	Summary
Davis-Besse	N/A	11/23/94	Broad	None	NRR staff audited the penetration seal program. On the basis of the audit, the staff concluded that the licensee had implemented and maintained an acceptable fire barrier penetration seal program and that no significant problems existed with the fire barrier penetration seal installations. The staff did not find information that suggested problems with generic implications.
Diablo Canyon 1/2	94-01	03/15/94	Broad	Minor	In 1994, the licensee found that certain fire barrier penetration seals may not have met the required 3-hour fire rating because damming boards were not installed on both sides of silicone foam seals. A walkdown of additional seals revealed about 100 representative silicone foam seals with missing damming boards. The licensee has established a corrective action program. The staff followed up on the licensee's activities during inspections in February 1994 and March 1995. Inspectors concluded that the licensee had taken appropriate corrective actions. The staff is continuing to follow the licensee's actions.
	94-07	04/94	Broad	None	LER 94-01, "Inadequate Fire Barrier Penetration Seals Due to Lack of Damming Boards," was closed by inspectors.
	94-18	08/15/94	Broad	Minor	During an inspection of fire barrier penetration seals, inspectors noticed a breached seal. The breach in the seal was the result of ongoing work and the licensee had appropriate compensatory measures in place for the breached seal.
	95-03	05/01/95	Broad	Minor	Inspectors reviewed the licensee's corrective actions for LER 94-001, which reported inadequate silicone foam fire barriers due to lack of damming boards. Inspectors found that the licensee's actions were appropriate, but the item remained open, as action was still ongoing.
	96-13	08/18/96	Broad	None	Inspectors closed LER 94-001 concerning inadequate fire barrier penetration seals due to lack of damming boards. Licensee undertook a 100% inspection of required seals to document all installed configurations. Inspectors concluded that the licensee's program would correct the seal deficiencies.
D.C. Cook 1/2	94-012	06/94	Broad	Minor	Inspector noted that inoperable fire barrier penetration gap seals were a major problem at the plant, but the licensee had begun an aggressive program to inspect 485 additional gap seals.
Duane Arnold	93-012	10/93	Narrow	Minor	Inspectors described problems licensee was experiencing regarding fire barrier penetration seals. A major problem was noted in this area in an LER in 1992. The licensee was in the process of a 100% inspection of seals to identify problems.
	93-16	10/01/93	Narrow	Significant	Violation was issued to the licensee based on the lack of action taken regarding degraded barriers between control room and cable spreading room.

Plant	Report	Date	Scope	Findings	Summary
Farley 1/2	88-27	11/03/88	Narrow	Minor	Inspectors found several unsealed penetrations during a plant walkdown. These discrepancies were quickly dispositioned and repaired.
	94-30	01/06/95	Narrow	None	Inspectors reviewed licensee actions regarding notification from a foam seal vendor that self-extinguish times for a certain lot of RTV foam were out of specification. The licensee found one penetration seal that was formed of the suspect foam. At the time of the inspection, the licensee had scheduled to replace the penetration seal.
	95-20	01/96	Narrow	None	The licensee discovered conduit penetrations through a fire barrier without an internal seal. A broad review of conduit penetrations revealed that there were 125 conduits (3/4" to 4" diameter) that did not appear to be properly sealed. All conduit inspections and repairs had been completed and documented.
	96-13	12/23/96	Narrow	None	Inspectors concluded that licensee's evaluation of IN 94-28, "Potential Problems With Fire Barrier Penetration Seals," was appropriate and required corrective actions were completed.
	97-12	09/26/97	Narrow	Minor	Inspectors reviewed silicone foam penetration seals. Seal documentation did not contain important design parameters. GL 86-10 evaluations were not available for identified deviations. IFI 97-12-01 was opened to track these discrepancies.
Fermi 2	94-012	11/23/94	Narrow	None	As part of a restart inspection, inspectors noted that the licensee had reviewed installation records, including QA/QC records, for all installed seals and found them indicative of proper installations. In addition, the licensee had not found any indications of improper installation upon removal and inspection of several penetration seals.
FitzPatrick	93-12	07/15/93	Broad	None	Inspectors reviewed licensee special report 93-003 regarding nonfunctional fire barrier penetration seal. Inspectors concluded that appropriate action was taken to address the event.
	93-14	08/24/93	Broad	Minor	A seal was opened as part of a plant modification and was not properly restored. The seal was inoperable for more than 7 days before it was repaired and returned to operability. Inspectors issued a non-cited violation due to the licensees prompt actions.
	93-26	01/04/94	Narrow	None	As part of a fire protection inspection, penetration seals were inspected.
Ginna	94-14	06/13/94	Broad	None	Inspector verified that evaluations for existing penetration seal materials supported their qualification for use throughout the plant. Inspector determined that qualification documentation for penetration seal materials was concise. Inspector concluded that controls for maintaining integrity of fire barriers were good and considered this a fire protection program strength.
Grand Gulf 1	90-10	06/04/90	Narrow	Minor	Inspectors reviewed an annual fire protection audit, which stated that a number of fire barrier penetrations that require repair or rework were identified during a walkdown of Unit 1 rated penetrations.

Plant	Report	Date	Scope	Findings	Summary
Haddam Neck	93-08	07/26/93	Narrow	None	Inspectors closed out LER 93-003, "Fire Barriers Inoperable Due to Fire Seal Deficiencies."
	95-09	06/19/95	Broad	None	Inspector reviewed the fire barrier and penetration seal program to verify the adequacy of seal installations, qualification, and surveillance activities. Inspector found that the licensee conducted a 100 % visual inspection as part of its seal upgrade program in 1988. Licensee found 20 degraded or inoperable seals since the upgrade program. Inspector concluded that the licensee took prompt and appropriate corrective actions. On the basis of the inspection, inspector concluded that no safety concerns exist at the facility regarding fire barriers.
Hatch 1/2	88-21	08/23/88	Narrow	None	Inspectors reviewed the licensee's actions taken in response to IN 88-04, "Inadequate Qualification and Documentation of Fire Barrier Penetration Seals." Inspectors concluded that the licensee had planned to implement an adequate action plan, and that the implementation would be the subject of a future inspection.
	91-30	12/19/91	Narrow	None	Inspectors noted that penetration seals were acceptable during a general plant walkdown.
	92-09	04/20/92	Narrow	None	Inspectors noted that penetration seals were acceptable during a general plant walkdown.
	93-22	11/2/93	Narrow	None	Inspectors noted that penetration seals were acceptable during a general plant walkdown.
	97-01	03/24/97	Narrow	None	Inspectors concluded that licensee's evaluation of IN 94-28, "Potential Problems With Fire Barrier Penetration Seals," was appropriate and required corrective actions were completed.
	97-03	6/17/97	Broad	None	Inspectors reviewed procedures, drawings, and other documents related to fire-rated sealed penetrations and conducted walkdowns of selected sealed penetrations. Inspectors concluded that the licensee's program for determining the operability of sealed penetrations was adequate. No deficiencies were identified with the penetrations inspected.
	98-01	04/21/98	Broad	Minor	Inspectors reviewed several fire barrier penetration seals, including supporting documentation. A visual inspection did not reveal any discrepancies. Inspectors opened IFI 98-01-05 to track issues related to fire test documentation that was unavailable at the time of the inspection.
Indian Point 2	93-18	09/13/93	Broad	None	Inspectors reviewed the licensee's fire barrier penetration seal installation and surveillance program and the licensee's actions in response to IN 88-04. No discrepancies were found. The licensee does not use silicone foam-type penetration seals. Grout seals are utilized.

Plant	Report	Date	Scope	Findings	Summary
Indian Point 3	93-24	12/14/93	Broad	Minor	Inspectors opened URI 93-24-03, which concerned operability determinations of degraded and potentially nonconforming fire barriers and fire barrier penetration seals and the methodology that the licensee used to determine self-ignition temperatures of cables installed in penetrations in the plant. The latter issue has yet to be resolved.
	93-80	06/21/93	Narrow	Minor	Inspectors identified weaknesses in programs dealing with fire barrier penetration seals. Specifically licensee commitments to revise technical specifications to add fire barrier penetrations needed to meet Section III.G of Appendix R.
	95-10	06/26/95	Narrow	Minor	Inspectors questioned the methodology used by the licensee to determine the self-ignition temperature of cables that pass through penetration seals. However, inspectors had found the licensee's penetration seal analyses and supporting documentation to be generally sufficient. The NRC is currently tracking corrective actions at IP3.
	95-81	05/11/95	Narrow	Minor	Inspectors reviewed fire barrier penetration seal qualification tests and concluded that insufficient evidence was available to support the cable ignition temperatures of cables installed at IP3. (Similar to preceding summary.)
Kewaunee	96-004	06/05/96	Broad	Minor	Inspector cited the licensee for a lack of corrective action in restoring a degraded fire barrier penetration seal that was identified as impaired, but not dispositioned or repaired. There were no compensatory measures taken for this degraded fire barrier. The licensee was issued a Level IV violation.
LaSalle	96-04	07/03/96	Narrow	Broad	Inspectors noted that barrier seals in the plant were in generally good condition.
Maine Yankee	95-15	09/20/95	Broad	None	Inspector reviewed the fire barrier program to verify the adequacy of penetration seal installations, qualification, and inspection activities. Inspector concluded that the licensee's procedures for seal inspections and training provided to seal inspectors were good for maintaining proper seal configuration and early detection of degraded conditions.
	96-08	09/16/96	Broad	Significant	Inspectors reviewed actions taken by the licensee to address problems identified with penetration seals. Inspectors concluded that the licensee took prompt and effective actions to address these problems.
	97-03	06/05/97	Broad	Minor	Inspectors reviewed the licensee's activities involving the fire barrier penetration seal repair project. 90% of the 2600 seals inspected were determined to require repair or replacement.
McGuire 1/2	89-03	04/06/89	Narrow	None	Inspectors reviewed the licensee's actions in response to IN 88-04 and found that they were adequate to address the concerns outlined in the IN.
	92-01	02/19/92	Narrow	None	Inspectors noted that penetration seals were acceptable during a general plant walkdown.
	98-07	08/04/98	Broad	None	Inspectors reviewed licensee's fire barrier penetration seal program and concluded that seal designs were properly supported by seal testing documentation, vendor data, design data and inspection.

Plant	Report	Date	Scope	Findings	Summary
Millstone 1/3 Millstone 2	93-19, 93-14, 93-15	10/06/93	Narrow	None	Inspectors reviewed licensee corrective actions for LER 93-06. 100% inspections are done every 18 months for Unit 1. Inspectors reviewed revised penetration seal surveillance procedure and found it adequate. Overall, the corrective actions were appropriate.
Monticello	92-007	04/10/92	Narrow	None	Inspector observed fire penetration seals while conducting a plant walkdown and did not observe problems.
	93-005	04/93	Narrow	None	Inspectors closed LER 91-21, which reported inoperable fire barrier penetration seals due to pipe movement caused by a water hammer. Inspectors felt the actions taken by the licensee to resolve this problem were adequate.
North Anna 1 North Anna 2	88-13	09/13/88	Narrow	None	Inspectors reviewed several exemptions requests and inspected penetration seals including supporting documentation. Inspectors did not identify any discrepancies.
	92-18	10/19/92	Broad	Significant	Inspectors identified several degraded penetration seals and upon review of the penetration seal program found deficiencies in procedures and documentation. Two violations were issued for failure to maintain penetration fire barriers (92-18-04) and failure to establish adequate fire barrier inspection procedure (92-18-05).
	93-13	03/30/93	Narrow	None	During a general plant walkdown, inspectors noted that penetration seals were acceptable.
	93-20	09/17/93	Broad	Minor	Inspectors observed penetration seal inspections where removal of marinite damming boards revealed gaps in penetration seal underneath. The same problems were found in 5 other seals. Fire watches were put into place until the inspections and repair were completed.
	94-10	06/09/94	Narrow	None	Inspectors reviewed licensee's corrective actions for violations 92-18-04 and 92-18-05. Violation 92-18-04 remained open, pending licensee's review of penetration seal inspection schedule. Violation 92-18-05 was closed.
	94-15	08/02/94	Broad	None	Inspectors reviewed the licensee's results from penetration seal inspections. Based on the conclusions of the inspections, Inspector closed violation 92-18-04.
	96-13	02/07/97	Broad	None	In 1995, the licensee initiated destructive inspections of penetration seals. It found and repaired a number of degraded seals. On the basis of this inspection, inspectors concluded that the licensee's corrective action program was very effective.

Plant	Report	Date	Scope	Findings	Summary
Oconee 1/2/3	88-19	07/21/88	Broad	None	Inspectors closed LER 88-05 on inoperable fire barrier penetration seals based on their review of the licensee's corrective actions.
	91-14	08/01/91	Narrow	None	Inspectors reviewed the licensee's procedure for 18-month surveillance of fire barrier penetration seals. They also inspected seals during a plant walkdown and noted no discrepancies.
	97-15	12/15/97	Broad	Minor	Inspectors reviewed the licensee's penetration seal program. The licensee had initiated a reverification program for penetration seals in all 3 units. Inspectors opened IFI 97-15-07 to follow this effort.
Oyster Creek	93-10	06/21/93	Broad	None	Inspectors viewed penetration seals during plant walkdown. No visible discrepancies were noted. Inspectors also reviewed licensee actions in response to IN 88-56. Licensee conducted inspections of installed silicone foam quality during installation and at periodic intervals by removing damming boards.
	95-11	07/21/95	Broad	Minor	Inspection was conducted because licensee reported finding degraded penetration seals (125 of about 1560 seals) during its 18-month seal inspection program. Inspector concluded that the licensee had accurately identified, evaluated, and initiated proper compensatory and/or repair activities. Inspector concluded that there were no outstanding operability or functionality issues.
Palisades	92-010	03/92	Narrow	None	Inspector reviewed licensee's fire barrier penetration surveillance procedure. No discrepancies were noted.
Palo Verde 1/2/3	94-29	09/02/94	Broad	None	Inspector reviewed the licensee's fire barrier seal program and found that extensive inspections had been completed and deficiencies were being addressed by the licensee.
Peach Bottom 2/3	93-09	05/14/93	Broad	Minor	Inspectors reviewed the licensee's fire barrier penetration seal installation and surveillance program. Voids were discovered in some silicone foam penetration seals. The licensee responded by inspecting all seals supported by a given detail. Inspectors concluded that the licensee's penetration repair program appeared to be an adequate approach for identifying and correcting nonconforming penetrations.
Perry 1	96-016	02/04/97	Narrow	Minor	Inspector opened an unresolved item regarding penetration seals that were installed in a different configuration from the supporting tested assembly. The licensee was to complete an engineering evaluation.
Pilgrim 1	92-27	12/30/92	Narrow	None	During a fire protection inspection, while on a plant tour, inspectors inspected penetration seals. No adverse conditions were noted.
	97-03	07/22/97	Broad	Minor	Inspector discovered a penetration seal with a small void at the top of the seal. The seal was determined to be degraded but operable. The seal was to be repaired.
Prairie Island 1/2	92-010	08/14/92	Narrow	None	Seals for separation of diesel generators from other plant areas were inspected and verified as 3-hour rated.

Plant	Report	Date	Scope	Findings	Summary
River Bend Station	94-17	01/17/95	Narrow	None	Inspection team observed penetration seals during a fire protection-related plant tour. No discrepancies were noted.
	94-22	01/26/95	Narrow	None	Inspectors questioned the radiation shielding capability of Kaowool installed as a penetration seal. The licensee was able to adequately justify the application.
	95-01	03/08/95	Narrow	Minor	Inspectors found that inadequate corrective actions for misapplication of seal material in 1991 caused seals to degraded by high ambient temperatures. Inspectors opened URI 95-01-02.
	95-02	05/03/95	Narrow	Minor	Inspector follow-up on URI 95-01-02 concluded that the licensee was acting appropriately, but more work was needed to resolve the problems.
	95-17	06/09/95	Narrow	Minor	The licensee received a non-cited violation for failure to promptly identify and correct the inadequate design of the boot seal that had degraded. Inspectors closed URI 95-01-02, based on the licensee's ongoing efforts to correct the seal problem.
H.B. Robinson 2	88-31	01/12/89	Narrow	Minor	Inspectors generated IFI 88-31-01 based on their review of GL 86-10 evaluations of seals that did not meet the technical specification surveillance acceptance criteria. Seals were dispositioned in engineering evaluations rather than being repaired.
	90-15	08/06/90	Narrow	None	Inspectors reviewed the licensee's fire barrier inspection project, which was initiated to ensure that all seals were operable per plant technical specifications. IFI 88-31-01 remained open pending the completion of this project.
	91-13	05/17/91	Narrow	None	Inspectors closed LER 90-10 on an inoperable penetration seal, and IFI 88-31-01 based on the completion of the licensee's penetration seal inspection project. Several seals were scheduled to be repaired because of the inspection project.
	96-12	12/16/96	Narrow	None	Inspectors noted that penetration seals were adequate during a general plant walkdown. Also, inspectors concluded that licensee's evaluation of IN 94-28, "Potential Problems With Fire Barrier Penetration Seals," was appropriate and required corrective actions were completed.

Plant	Report	Date	Scope	Findings	Summary
St. Lucie 1 **St. Lucie 2**	96-08	07/08/96	Narrow	None	Inspectors evaluated the licensee's actions to resolve fire protection discrepancies during the 1996 Unit 1 refueling outage. The licensee had inspected penetration seals and found small cracks in the surfaces of the seals. Inspectors concluded that the discrepancies did not appear to degrade the fire resistance of the seals. However, the licensee considers seals with even cosmetic problems to be inoperable. Inspectors found that the licensee's corrective actions and compensatory measures were appropriate.
	97-06	08/25/97	Broad	Minor	Inspectors cited licensee for failure to promptly take appropriate corrective actions to resolve mechanical penetration seal deficiencies.
Salem 1 **Salem 2**	93-80	10/14/93	Narrow	None	Inspectors reviewed results of 18-month fire barrier penetration seal surveillance conducted by the licensee. No discrepancies were noted.
	96-01	03/25/96	Broad	None	Inspectors reviewed the licensee's procedure for fire barrier penetration seal inspections.
	96-10	10/30/96	Broad	None	This issue was a restart action plan item. Inspectors reviewed work done during the penetration seal improvement program and concluded that the quality and configuration of penetration seals were acceptable.
	97-09	06/03/97	Broad	None	Inspectors reviewed the qualification-type fire endurance tests and associated engineering evaluations for certain seal designs in floors and walls in Unit 1 and Unit 2 auxiliary buildings. Inspectors focused on verifying that design and installation parameters for the as-built configurations were bounded and justified by the licensee's engineering evaluations. Inspectors concluded that the licensee's engineering analysis methods appeared to have established a basis that the as-built seal designs would accomplish their intended function.
San Onofre 2/3	94-01	01/28/94	Broad	None	The licensee conducted a 100 % reverification program of the installed configurations as a part of the validation of the Plant and Equipment Data Management System database. The licensee found that 4 of 1500 seals (a 20 % sample of a total of 7000 seals) did not meet acceptance criteria. (The reverification process was ongoing at the time of the inspection.) Inspector walked down and verified the adequacy of a sample of installed seals. Inspector did not report any safety-significant problems.

Plant	Report	Date	Scope	Findings	Summary
Sequoyah 1/2	88-54	01/13/89	Narrow	Minor	Inspectors reviewed procedures for licensee penetration seal inspections. Inspectors also found 2 fire barrier penetration seals that were breached by a rubber hose. Inspectors opened IFI 88-54-05.
	92-14	06/05/92	Broad	Minor	Inspectors closed LERs 91-010, 91-008, 91-016, and 91-012. Licensee had planned a 100% inspection and reverification of all installed seals in accordance with IN 88-04.
	94-16	07/19/94	Broad	None	Inspectors reviewed the licensee's response to IN 88-04, which included inspection and seal re-work. Inspectors concluded that the licensee's followup on the IN was adequate.
	96-02	04/22/96	Broad	Minor	Inspectors reviewed a 1994 licensee audit in which items identified included inadequate design control over fire barrier penetration seals and restoration of pen seals to operability following maintenance. Corrective actions on these items were incomplete at the time of the inspection.
	96-10	09/27/96	Broad	Minor	Inspectors reported that a 100% seal inspection had been completed (24,500 seals inspected) and 1500 seals with design documentation problems remained to be resolved. Scheduled for completion late 1997.
	97-03	05/12/97	Narrow	None	Inspectors concluded that licensee's evaluation of IN 94-28, "Potential Problems With Fire Barrier Penetration Seals," was appropriate and required corrective actions were completed.
	98-07	08/07/98	Broad	Minor	Inspectors reviewed licensee's penetration seal program, including a walkdown of 24 silicone foam penetration seals. Inspectors opened IFI 97-08-10 to track licensee's actions concerning the evaluation of installed seal configurations that are not adequately supported by a fire test.
Shearon Harris	95-02	03/02/95	Narrow	None	Inspectors observed penetration seal 18-month visual inspection conducted by licensee personnel. Performance of the inspection was found to be satisfactory.
	98-01	03/27/98	Broad	Minor	Inspectors reviewed the licensee's penetration seal program. 3 of the seals inspected lacked adequate supporting documentation and engineering analysis. A violation was cited based on this weakness
South Texas 1/2	94-15	06/07/94	Narrow	Minor	IFI regarding excessive shrinkage of penetration seals was closed in the report.
	95-01	03/06/95	Narrow	None	Inspector visually inspected penetration seals in various fire areas and found no discrepancies.

Plant	Report	Date	Scope	Findings	Summary
Surry 1/2	88-07	03/17/88	Narrow	None	During a fire protection inspection, inspectors reviewed procedures for the licensee's fire stop and fire retardant coatings surveillances.
	93-18	07/27/93	Narrow	None	During a fire protection inspection, inspectors reviewed procedures for the licensee's fire stop and fire retardant coatings surveillances.
	96-10	10/28/96	Narrow	None	Inspectors concluded that licensee's evaluation of IN 94-28, "Potential Problems With Fire Barrier Penetration Seals," was appropriate and required corrective actions were completed.
Susquehanna 1	95-12	08/02/95	Narrow	Minor	Inspectors followed up on LERs 94-003 and 94-007 for a missing seal and a degraded seal. Both discrepancies were corrected.
	95-14	07/31/95	Broad	None	Inspector conducted a comprehensive inspection of the licensee's penetration seal program including, reviewing the adequacy of the penetration seal installations, qualification, and inspection activities. Inspector also assessed the appropriateness of acceptance criteria for validating operability and degradation. Inspector concluded that the licensee had an excellent program.
	96-201	04/05/96	Broad	Minor	NRR staff inspected the fire barrier penetration seal program and found the damming material missing from one penetration seal. The licensee took immediate corrective actions. Inspectors concluded that the licensee had implemented and maintained an acceptable fire barrier penetration seal program. Inspectors did not find safety-significant problems or evidence of generic problems with penetration seals.
Turkey Point 3/4	88-37	01/05/89	Narrow	Minor	During a fire protection inspection, inspector reviewed the procedure for penetration fire barrier surveillances. Inspector noted that the procedure did not identify that all the installed fire barriers met Appendix R requirements. The procedure was being revised at the time of the inspection.
	92-23	10/29/92	Narrow	None	During a fire protection inspection, inspector reviewed the penetration seal inspection procedure. Inspector noted no discrepancies in penetration seals during a plant walk-down.
	96-06	06/03/96	Narrow	Minor	Licensee QA audits of fire protection program were reviewed. Findings regarding penetration seal documentation were identified. Corrective actions were determined to be adequate.
	97-11	11/24/97	Broad	Minor	Inspectors reviewed the licensee's actions in response to INs 88-04, 88-56, and 94-28. The licensee was evaluating the adequacy of silicone elastomer seals and found some seals without supporting documentation. Inspectors opened IFI 97-11-04 to track the licensee's progress in evaluating all the seals installed at the plant.

Plant	Report	Date	Scope	Findings	Summary
Vermont Yankee	93-05	05/13/93	Broad	Significant	A violation was issued to the licensee for inadequate application of quality principles to the original installation and the subsequent ineffective periodic inspections of the fire barrier penetration seals installed in the reactor building, control building, and diesel generator rooms.
Virgil C. Summer	96-11	11/25/96	Narrow	None	Inspectors concluded that licensee's evaluation of IN 94-28, "Potential Problems with Fire Barrier Penetration Seals," was appropriate and required corrective actions were completed.
Vogtle 1/2	88-24	06/29/88	Narrow	Minor	Following a spurious actuation of a fire suppression system, several fire penetration seals allowed the passage of water from one fire area to another. Inspector issued a Level IV violation for the failure to adequately design and install watertight penetration seals.
	91-10	06/13/91	Narrow	Minor	During a fire protection inspection, inspector (during a walkdown), found an unsealed penetration. Inspector issued a Level IV violation for this and other fire protection deficiencies.
	92-13	08/04/92	Broad	None	Inspectors completed an inspection on fire barrier penetration seals. Inspectors reviewed surveillances, noted discrepancies, and confirmed that all deficiencies were corrected. There were no findings in this area.
	93-08	05/17/93	Narrow	None	Inspectors found penetration seals to be adequate during a general plant walkdown. Violation 91-10-01, concerning corrective actions for missing penetration seals, was closed.
	95-31	02/96	Narrow	None	Inspectors closed LER 95-01 for lack of penetration seals placing plant in condition outside of design basis. Corrective actions were found to be adequate.
	97-01	04/14/97	Narrow	None	Inspectors concluded that licensee's evaluation of IN 94-28, "Potential Problems With Fire Barrier Penetration Seals," was appropriate and required corrective actions were completed.
	97-12	02/23/98	Broad	None	Inspectors reviewed the licensee's fire barrier penetration seals program, which included the inspection of individual penetration seals. Inspectors concluded that seal designs were properly supported by testing documentation, vendor data, installer qualification and training records, and quality assurance inspection records.

Plant	Report	Date	Scope	Findings	Summary
Washington Nuclear 2	94-08	02/25/94	Broad	Minor	In December 1993, the licensee began a review of issues related to its penetration seal inspection program. The licensee found deficiencies with original installations, periodic inspections, and repairs. The licensee declared all seals inoperable, established compensatory measures, and initiated a comprehensive penetration seal upgrade program. NRC Region IV conducted three inspections of the program. Inspectors concluded that the licensee was taking aggressive corrective actions.
	94-09	05/04/94	Broad	None	SALP report noted that penetration seal problems were not properly addressed by the licensee until the NRC became involved.
	94-28	11/09/94	Broad	Significant	The staff issued a violation (94-28-01) for not taking prompt compensatory measures upon the discovery of installation and inspection deficiencies for fire barrier penetration seals. Inspectors viewed approximately 100 penetration seals and noted that many had small cracks or gaps along the seal-wall interface. Inspectors did not believe that the deficiencies made the barriers nonfunctional.
	95-18	06/29/95	Broad	None	Inspectors closed violation 94-28-01. The licensee had completely restructured its fire protection program, including its penetration seal program.
	95-201	10/03/95	Broad	None	An NRC integrated assessment team inspected the licensee activities mentioned in the previous inspection reports. The team assessed licensee effectiveness in identifying issues, performing root cause analyses, and implementing corrective actions. The inspection focused on the areas of maintenance and engineering. The team inspected activities involving procurement, storage, installation, quality control, and long-term maintenance associated with the installation and maintenance of penetration seals. The team concluded that the licensee's current performance in the areas of receipt inspection and storage control, quality control, and inspection and surveillance was adequate. The assessment team also considered the licensee's corrective action program on penetration seals to be a strength.
Waterford 3	N/A	10/07/94	Broad	None	NRR staff audited the penetration seal program. The staff found several minor weaknesses with fire test results and training records. The staff concluded, however, that the fire barrier penetration seal program was satisfactory and that the discrepancies did not create any problems with the penetration seal installations. The staff did not find safety-significant problems or evidence to suggest that generic problems existed with penetration seals.
	95-11	02/16/95	Narrow	None	Inspector visually inspected penetration seals in various fire areas. No discrepancies were identified. Fire barrier penetration seal program implementing procedure was also reviewed.

Plant	Report	Date	Scope	Findings	Summary
Watts Bar 1	94-62	11/16/94	Narrow	Minor	Inspectors opened construction deficiency reports (CDRs) 85-18/19 and 90-10 for fire-rated penetration deficiencies and unqualified cable penetration seals. These issues were inspected several times over a 2-year period.
	94-78	12/21/94	Broad	None	Inspectors reviewed penetrations and supporting documentation for a number of seals. In addition, inspectors observed several seal installations. Inspectors concluded that an effective program was being implemented for the evaluation of existing electrical and mechanical fire barrier penetration seals and the repair, modification, and installation of penetration seals to meet design requirements.
	95-32	06/09/95	Narrow	None	Inspectors continued followup on CDR 85-19.
	95-38	07/11/95	Narrow	Minor	Inspectors discovered degraded penetration seals during a plant tour.
	95-39	07/18/95	Narrow	None	Inspectors closed CDR 87-13, which concerned deficiencies with mechanical fire protection penetration seals.
	95-40	09/12/95	Narrow	Minor	Documentation of the fire testing for fire barrier penetration seals did not conform to the design details for some installed seals. This follows from previous CDRs.
	95-45	08/15/95	Narrow	Minor	Inspector noted that a penetration seal had been breached.
	95-68	10/19/95	Broad	None	Inspectors reviewed design details and QA/QC records, and walked down penetration seals. No discrepancies were identified for the seals that were reviewed. During the walkdown some seals were noticed to have damaged damming boards. The applicant was already aware of these deficiencies.
	95-72	11/17/95	Narrow	None	Inspectors closed CDR 85-19 for penetration assembly deficiencies.
	95-77	12/06/95	Narrow	None	Inspectors closed second CDR (90-10) for unqualified penetration seals.
Wolf Creek	94-02	04/15/94	Broad	Minor	Cracks found in fire barrier material which formed a penetration seal between two areas.
	95-19	08/10/95	Narrow	None	Inspector visually inspected penetration seals in various fire areas. No discrepancies were identified.

Appendix J

Plants Known To Have Performed
100-Percent Penetration Seal Inspections

Appendix R plants (plants operating prior to January 1, 1979) shown in **bold** font.

Plant	Comm. Op.	Reference
Arkansas Nuclear One 1	1974	LER 91-016.
Arkansas Nuclear One 2	1980	LER 91-016
Big Rock Point	1963	LERs 89-006 and 91-001.
Browns Ferry 2/3	1975/1977	NRC IR 98-01.
Brunswick 1/2	1977/1975	LER 93-006-00.
Calvert Cliffs 1/2	1975/1977	NRC IR 94-15.
Catawba 1/2	1985/1986	McGuire LER 88-030-01.
D.C. Cook 1/2	1975/1978	LER 96-004-00.
Diablo Canyon 1/2	1985/1986	LER 94-001-00. NRC IRs 94-01 and 95-03.
Duane Arnold	1975	NRC IR 93-012.
FitzPatrick	1975	LER 91-024-01.
Fort Calhoun	1973	LER 90-022.
Haddam Neck	1968	LERs 89-001-00 and 95-001-00. NRC IR 95-09.
Indian Point 3	1976	NRC IR 95-81.
McGuire 1/2	1981/1984	LER 88-030-01.
Maine Yankee	1973	LER 96-017-00. NRC IR 96-08.
Millstone 1	1986	LER 93-006-01, NRC IR 93-19.
Monticello	1971	LER 89-001-00.
Nine Mile Point 1	1969	LER 88-009-00.
Oconee 1/2/3	1973/1973/1974	LERs 89-010-03 and 88-005, NRC IR 97-15.
Palo Verde 1/2/3	1986/1986/1988	Letter of March 16, 1990.
River Bend Station	1986	LER 89-010-03.
H.B. Robinson 2	1971	LER 91-010-01.
Salem 1	1977	LER 88-013-00.
Salem 2	1981	LER 88-013-00.

COMM. OP. = Date of Commercial Operation, **LER** = Licensee Event Report, **NRC IR** = NRC Inspection Report

Plant	Comm. Op.	Reference
San Onofre 2/3	1982/1983	NRC IR 94-01.
Susquehanna 1	1983	NRC IR 95-12.
Sequoyah 1/2	1981/1982	NRC IR 96-10.
Vermont Yankee	1972	LER 93-001.
Washington Nuclear 2	1984	LER 88-008-00. NRC IRs 94-08, 94-28, 95-18, and 95-201.
Watts Bar 1	1996	NRC IR 95-77.

Appendix K
Reference Summary

Appendix R plants (plants operating prior to January 1, 1979) shown in **bold** font.

Plant	Comm. Op.	Reference
Arkansas Nuclear One 1	1974	LERs 89-003-00, 90-004-00, 90-04-01, 90-004-02, 90-017-00, and 90-023-00.
Arkansas Nuclear One 2	1980	LERs 87-001-00 and 91-016-00.
Beaver Valley 1	1976	NRC IRs 93-12 and 93-13.
Beaver Valley 2	1987	NRC IRs 93-12 and 93-13.
Big Rock Point	1963	LERs 89-006-00, 89-006-01, 91-001-00, and 91-001-01.
Braidwood 1/2	1988/1988	N/A.
Browns Ferry 1/2/3	1974/1975/1977	NRC IRs 89-28, 90-11, 92-11, 95-60, and 98-01.
Brunswick 1/2	1977/1975	LER 93-006-00; NRC IRs 92-31, 93-08, 93-38, 97-07, and 97-13.
Byron 1/2	1985/1987	NRC IR 92-007.
Callaway	1984	NRC IR 94-12.
Calvert Cliffs 1/2	1975/1977	LERs 89-002-00, 89-002-01, and 95-004-00; NRC IRs 94-15, 93-99, 95-08, and 96-201.
Catawba 1/2	1985/1986	McGuire LER 88-030-01; NRC IRs 91-22, 97-07, and 98-07.
Clinton	1987	LERs 89-006-00 and 98-021-00.
Comanche Peak 1/2	1990/1993	NRC IRs 96-10 and 96-12.
Cooper	1974	LER 94-008-00; NRC IR 95-17.
Crystal River 3	1977	NRC IRs 92-18 and 97-18.
Davis-Besse	1978	1994 NRR audit.
D.C. Cook 1/2	1975/1978	LER 96-004-00; NRC IR 94-012.
Diablo Canyon 1/2	1985/1986	LERs 94-001-00, 94-001-01, 96-011-00, and 96-011-01; NRC IRs 94-01, 94-07, 94-18, 95-03, and 96-13.
Dresden 2/3	1970/1971	LER 89-030-00.
Duane Arnold	1975	LERs 92-003-00, 92-007-00, and 92-007-01; NRC IRs 93-012 and 93-16.
Farley 1	1977	NRC IRs 88-27, 94-30, 95-20, 96-13, and 97-12.
Farley 2	1981	NRC IRs 88-27, 94-30, 95-20, 96-13, and 97-12.

COMM. OP. = Date of Commercial Operation, **LER** = License Event Report, **NRC IR** = NRC Inspection Report

Plant	Comm. Op.	Reference
Fermi 2	1988	LERs 97-014-00 and 97-014-01; NRC IR 94-012.
FitzPatrick	1975	LERs 87-011-00, 87-011-01, 91-024-00, and 91-024-01; NRC IRs 93-12, 93-14, and 93-26.
Fort Calhoun	1973	LERs 90-022-00, 90-022-01, and 90-022-02.
Fort St. Vrain		LERs 87-006-00. 87-006-01, 89-014-00, and 89-014-01.
Ginna	1970	LER 88-009-00; NRC IR 94-14.
Grand Gulf 1	1985	NRC IR 90-10.
Haddam Neck	1968	LERs 89-001-00, 89-001-01, 92-008-00, 93-003-00, 95-001-00, and 95-001-01; NRC IRs 93-08 and 95-09.
Hatch 1	1975	NRC IRs 88-21, 91-30, 92-09, 93-22, 97-01, 97-03, and 98-01.
Hatch 2	1979	NRC IRs 88-21, 91-30, 92-09, 93-22, 97-01, 97-03, and 98-01.
Hope Creek 1	1986	N/A.
Indian Point 2	1974	NRC IR 93-18.
Indian Point 3	1976	LER 93-029-00; NRC IRs 93-24, 93-80, 95-10, and 95-81.
Kewaunee	1974	NRC IR 96-004.
LaSalle 1/2	1984/1984	LER 93-009-00; NRC IR 96-04.
Limerick 1/2	1986/1990	N/A.
Maine Yankee	1973	LERs 94-010-00, 94-010-01, 96-017-00 and 97-017-01; NRC IRs 95-15 and 96-08.
McGuire 1/2	1981/1984	LERs 88-030-00 and 88-030-01; NRC IRs 89-03, 92-01, 97-03, and 98-07.
Millstone 1	1986	LERs 93-006-00 and 93-006-01; NRC IR 93-19.
Millstone 2	1975	LER 94-035-00; NRC IR 93-14.
Millstone 3	1986	NRC IR 93-15.
Monticello	1971	LERs 87-011-00, 89-001-00, 89-013-00, 89-013-01, 90-009-00, and 91-021-00; NRC IRs 92-007 and 93-005.
Nine Mile Point 1	1969	LERs 88-009-00, 88-009-01, and 88-009-02.
Nine Mile Point 2	1988	LERs 87-016-00, 87-016-01, and 87-018-00
North Anna 1	1978	LERs 88-007-00 and 89-003-00; NRC IRs 88-13, 92-18, 93-13, 93-20, 94-10, 94-15, and 96-13.
North Anna 2	1980	LERs 88-007-00 and 89-003-00; NRC IRs 88-13, 92-18, 93-13, 93-20, 94-10, 94-15, and 96-13.

Plant	Comm. Op.	Reference
Oconee 1/2/3	1973/1974/1974	LERs 88-005-00 and 89-010-03; NRC IRs 88-19, 91-14, and 97-15.
Oyster Creek	1969	NRC IRs 93-10 and 95-11.
Palisades	1971	LERs 89-024-00 and 96-009-00; NRC IR 92-010.
Palo Verde 1/2/3	1986/1986/1988	LERs 90-009-00 and 90-009-01; NRC IR 94-29
Peach Bottom 2/3	1974/1974	LER 91-013-00; NRC IR 93-09.
Perry 1	1987	NRC IR 96-06.
Pilgrim 1	1972	NRC IRs 92-27 and 97-03.
Point Beach 1/2	1970/1972	LER 91-007-00.
Prairie Island 1/2	1973/1974	LER 98-003-00; NRC IR 92-010.
Quad Cities 1/2	1973/1973	LER 87-028-00
River Bend Station	1986	LERs 87-021-00, 88-009-00, 88-009-01, 88-009-02, 89-005-00, 89-010-00, 89-010-01, 89-010-02, 89-010-03, 89-010-04, and 89-010-05; NRC IRs 94-17, 94-22, 95-01, 95-02, and 95-17.
H.B. Robinson 2	1971	LERs 88-018-00, 88-018-01, 90-003-00, 90-008-00, 90-010-00, 90-010-01, and 91-010-01; NRC IRs 88-31, 90-15, 91-13, and 96-12.
St. Lucie 1	1976	LERs 97-004-00 and 97-008-00; NRC IRs 96-08 and 97-06.
St. Lucie 2	1983	LERs 97-004-00 and 97-008-00; NRC IRs 96-08 and 97-06.
Salem 1	1977	LERs 87-007-00, 88-013-00, and 88-014-00; NRC IRs 93-80, 96-01, 96-10, and 97-09.
Salem 2	1981	LERs 87-007-00, 88-013-00, and 88-014-00; NRC IRs 93-80, 96-01, 96-10, and 97-09.
San Onofre 2/3	1983/1984	NRC IR 94-01.
Seabrook 1	1990	LERs 89-011-00 and 89-011-01
Sequoyah 1/2	1981/1982	LERs 91-013-00, 91-013-01, 91-016-00, and 91-016-01; NRC IRs 88-54, 92-14, 94-16, 96-02, 96-10, 97-03, and 98-07.
Shearon Harris	1987	NRC IRs 95-02 and 98-01.
South Texas 1/2	1988/1989	NRC IRs 94-15 and 95-01.
Susquehanna 1	1983	LERs 87-011-00, 89-019-00, and 95-011-00; NRC IRs 95-12, 95-14, and 96-201.
Summer	1984	NRC IR 96-11.
Surry 1/2	1972/1973	NRC IRs 88-07, 93-18, and 96-10.

Plant	Comm. Op.	Reference
Three Mile Island 1	1974	LER 87-003-00
Trojan		LERs 90-022-00, 90-022-01, 92-006-00, 92-006-01, 92-011-00, 92-026-00, 92-026-01, 92-026-02, 92-026-03, 92-026-04, 92-026-05; 92-031-00; 92-034-00, 93-001-00, and 93-002-00.
Turkey Point 3/4	1972/1973	NRC IRs 88-37, 92-23, 96-06, and 97-11.
Vermont Yankee	1972	LERs 93-001-00, 93-001-01, 93-001-02, 94-018-00, 94-018-01, 95-004-00, 96-026-00, 96-026-01, 98-001-00, 98-001-01, 98-008-00, 98-008-01, 98-014-00, 98-014-01; NRC IR 93-05.
Vogtle 1/2	1987/1989	NRC IRs 88-24, 91-10, 92-13, 93-08, 95-31, 97-01, and 97-12.
Washington Nuclear 2	1984	LERs 87-004-00, 87-029-00, 87-030-00, 88-008-00, 88-008-01, 94-002-00, and 94-002-01; NRC IRs 94-08, 94-09, 94-28, 95-18, and 95-201.
Waterford 3	1985	LERs 88-011-00, 88-025-00, 88-030-00, 88-030-01, 88-030-02, 88-030-03, 90-019-00, 90-019-01, and 90-019-02; NRC IR 95-11 and 1994 NRR audit.
Watts Bar 1	1996	NRC IRs 94-62, 94-78, 95-32, 95-38, 95-39, 95-40, 95-45, 95-68, 95-72, and 95-77.
Wolf Creek 1	1985	LERs 87-001-00, 87-010-00, 87-010-01, and 87-010-02; NRC IRs 94-02 and 95-19.
Zion 1/2	1973/1974	N/A.

NRC FORM 335 (2-89) NRCM 1102, 3201, 3202	U.S. NUCLEAR REGULATORY COMMISSION **BIBLIOGRAPHIC DATA SHEET** (See instructions on the reverse)	1. REPORT NUMBER (Assigned by NRC, Add Vol., Supp., Rev., and Addendum Numbers, if any.) NUREG – 1552, Supplement 1

2. TITLE AND SUBTITLE

Fire Barrier Penetration Seals in Nuclear Power Plants

3. DATE REPORT PUBLISHED	
MONTH	YEAR
January	1999

4. FIN OR GRANT NUMBER

5. AUTHOR(S)

C. S. Bajwa and K. S. West

6. TYPE OF REPORT

Technical

7. PERIOD COVERED (Inclusive Dates)

8. PERFORMING ORGANIZATION – NAME AND ADDRESS (If NRC, provide Division, Office or Region, U.S. Nuclear Regulatory Commission, and mailing address; if contractor, provide name and mailing address.)

Division of Systems Safety Analysis
Office of Nuclear Reactor Regulation
U.S. Nuclear Regulatory Commission
Washington, DC 20555–0001

9. SPONSORING ORGANIZATION – NAME AND ADDRESS (If NRC, type "Same as above"; if contractor, provide NRC Division, Office or Region, U.S. Nuclear Regulatory Commission, and mailing address.)

Same as above

10. SUPPLEMENTARY NOTES

11. ABSTRACT (200 words or less)

Nuclear power plants use the "defense in depth" concept of echelons of fire protection to achieve a high degree of fire safety. The objective of this concept is to (1) prevent fires from starting; (2) rapidly detect, control, and extinguish those fires that do occur; and (3) protect structures, systems, and components important to safety so that a fire that is not promptly extinguished will not prevent the safe shutdown of the plant. Fire barrier penetration seals, which are but one part of one element of the fire protection defense-in-depth concept, are designed to maintain the fire rating of a barrier where penetrating items pass through the barrier. This is essential if the barrier is to confine a fire to the area in which it started or to protect plant systems and components within an area from a fire outside the area. On the basis of everything it found and considered, it is the staff's judgment that, overall, the issue of potential fire barrier penetration seal deficiencies does not affect safety. For the reasons given in this paper, typical penetration seal deficiencies do not necessarily equate to a lack of adequate protection or result in undue risk to public health and safety. It is the staff's opinion that continued licensee upkeep of existing penetration seal programs and continued NRC inspections are adequate (1) to ensure that penetration seal problems are discovered and resolved and (2) to maintain public health and safety.

12. KEY WORDS/DESCRIPTORS (List words or phrases that will assist researchers in locating the report.)

Fire Barrier Penetration Seals
Silicone Foam
Fire Barriers

13. AVAILABILITY STATEMENT
Unlimited

14. SECURITY CLASSIFICATION
(This Page)
Unclassified
(This Report)
Unclassified

15. NUMBER OF PAGES

16. PRICE

NRC FORM 335 (2–89)

www.ingramcontent.com/pod-product-compliance
Lightning Source LLC
Chambersburg PA
CBHW081602170526
45166CB00009B/2787